Springer Theses

Recognizing Outstanding Ph.D. Research

Aims and Scope

The series "Springer Theses" brings together a selection of the very best Ph.D. theses from around the world and across the physical sciences. Nominated and endorsed by two recognized specialists, each published volume has been selected for its scientific excellence and the high impact of its contents for the pertinent field of research. For greater accessibility to non-specialists, the published versions include an extended introduction, as well as a foreword by the student's supervisor explaining the special relevance of the work for the field. As a whole, the series will provide a valuable resource both for newcomers to the research fields described, and for other scientists seeking detailed background information on special questions. Finally, it provides an accredited documentation of the valuable contributions made by today's younger generation of scientists.

Theses may be nominated for publication in this series by heads of department at internationally leading universities or institutes and should fulfill all of the following criteria

- They must be written in good English.
- The topic should fall within the confines of Chemistry, Physics, Earth Sciences, Engineering and related interdisciplinary fields such as Materials, Nanoscience, Chemical Engineering, Complex Systems and Biophysics.
- The work reported in the thesis must represent a significant scientific advance.
- If the thesis includes previously published material, permission to reproduce this must be gained from the respective copyright holder (a maximum 30% of the thesis should be a verbatim reproduction from the author's previous publications).
- They must have been examined and passed during the 12 months prior to nomination.
- Each thesis should include a foreword by the supervisor outlining the significance of its content.
- The theses should have a clearly defined structure including an introduction accessible to new PhD students and scientists not expert in the relevant field.
- Indexed by zbMATH.

More information about this series at http://www.springer.com/series/8790

Lesya Horyn

A Search for Displaced Leptons in the ATLAS Detector

Doctoral Thesis accepted by the University of Chicago, USA

Springer

Lesya Horyn
Fermi National Accelerator Laboratory
Batavia, IL, USA

ISSN 2190-5053 ISSN 2190-5061 (electronic)
Springer Theses
ISBN 978-3-030-91671-8 ISBN 978-3-030-91672-5 (eBook)
https://doi.org/10.1007/978-3-030-91672-5

This Springer imprint is published by the registered company Springer Nature Switzerland AG
The registered company address is: Gewerbestrasse 11, 6330 Cham, Switzerland

To my father, who taught me to look to the stars, and to my mother, who taught me to look in a book.

Supervisor's Foreword

The Standard Model of Particle Physics is the current best and most successful theory to describe the building blocks of the universe. It explains how quarks (which make up protons and neutrons) and leptons (including charged leptons such as electrons and muons and neutral leptons such as neutrinos) make up all known matter. It also explains three of the four fundamental forces that govern the universe: electromagnetism, the strong force, and the weak force. Despite its success at explaining the universe, the Standard Model does have limits and is not a complete theory. For example, it does not explain the fourth fundamental force, gravity. The Higgs boson gives mass to quarks, charged leptons, and carriers of the weak force, but we do not yet know whether it also gives mass to neutrinos that interact very rarely with other matter in the universe. The Big Bang should have created equal amounts of matter and antimatter in the early universe. But today, everything we see is made almost entirely of matter. The Standard Model cannot explain what happened to the antimatter, or why we see a huge asymmetry between matter and antimatter. Also, about 95% of the universe is not made of ordinary matter as we know it. Instead, much of the universe consists of dark matter and dark energy that do not fit into the Standard Model. In order to unravel these deepest mysteries, new physics beyond the Standard Model (BSM) is required. Particle physicists have been searching for clues as to what might lie beyond the Standard Model.

The Large Hadron Collider (LHC) provides an outstanding opportunity to discover new physics or new particles predicted by BSM theories. So far, particle physicists from the ATLAS and CMS experiments at the LHC have ruled out all the most obvious candidates for new particles. However, nearly all of these searches have been based on an expectation that BSM particles decay promptly in our detector.

When long-lived particles decay, the tracks of their decay products are displaced from the primary collision point. Particles with displacement beyond a few millimeters are explicitly rejected by most searches at the LHC, in the interest of cutting down on backgrounds. This means that most new physics exclusions from LHC experiments don't apply to BSM particles with macroscopic lifetimes.

Lesya Horyn's thesis is on searching for long-lived new particles who travel beyond a few millimeters before decaying (macroscopic lifetimes) using the ATLAS detector. These particles arise naturally from many BSM theories, but have not been actively searched for in the past because ATLAS's physical structure, trigger system, reconstruction software, and analysis tools have all been developed with an assumption that BSM particles decay promptly in the detector. Searching for new physics outside the prompt paradigm, thus, requires a reinvention of techniques at every level.

More specifically, Lesya's work was on searching for displaced lepton pairs produced by decays of the slepton, the lepton's supersymmetric partner, in a Gauge-Mediated Supersymmetry model. In this new physics model, the slepton decays to a lepton and a gravitino, the graviton's supersymmetric partner. This process, however, is suppressed by the Planck scale, an energy scale around 1019 GeV, giving the slepton a macroscopic lifetime. The gravitino is invisible to the ATLAS detector, so the only visible signature is the displaced leptons resulting from the slepton decays. This analysis is a challenge, as its backgrounds are unique, requiring the invention of new methods to estimate them. Moreover, every detail of the lepton reconstruction software must be combed through, and re-tuned to optimize for displaced objects. This signature hadn't been explored at the LHC before this thesis. This work therefore provided a major leap forward in signal sensitivity and opened up new space to search for long-lived new particles.

Chicago, IL, USA
August 2021

Young-Kee Kim

Acknowledgments

I think it's safe to say that nothing about this PhD has gone the way I expected. I didn't expect to write this dissertation during a pandemic, and I didn't expect to feel so content at the end of this time. I came to the University of Chicago with a lot of self-doubt and not a lot of certainty, but I leave feeling like a scientist. For that feeling, I have to thank everyone who has guided and supported me over the last 6 years.

I've had the immense privilege of having some incredible mentors. Young-Kee, thank you for pushing me and always making sure I had the support I needed. I've learned so much from your perspective on the world and on physics. Tova, I can only hope that I can be as great of a postdoc to someone as you have been to me. Working with you has really shaped my identity as a scientist. Thank you for always treating me with respect. I will really miss working with you. Max, I truly wouldn't have made it through my first year without you. You've been such a generous mentor, completely independent of how closely we were working together at the time. Thank you for always taking the time to answer all my questions.

I've also learned that working with good people makes all the difference. The displaced leptons analysis was a small and scrappy team who all taught me a lot. Thank you to Elodie, for pushing this through until the very end, to Xiaohe for being such a wonderful undergrad to work with, and to Kate for being the most helpful at the most important time. Though not the focus of this thesis, I spent 4 years working on FTK with a truly incredible team. Thanks to Mel, Patrick, and Rui, for your guidance and opinions. And to Todd, my favorite physics younger brother, somehow we made it through together and I'm really grateful for your friendship. Liza, I've learned so much from you about problem solving, care, thoughtful work, and, of course, team building. Also, to Tyler, Stany, Bri, Jon, and the rest of the team that made a stressful project a lot of fun.

I've been extremely fortunate to have friends that always made me feel like I was at home, no matter where in the world I was. I'll always cherish my first year in Chicago, I was very lucky to have a wonderful cohort full of smart, caring people, especially Gautam, Mark, and Dani. But most especially Lipi, the best long distance friend and spontaneous roommate turned family. Thank you to my CERN lunch crew turned pandemic survival team, Max, Tova, Kate, Larry, Liza, Karri, Ann, Emma, and Chris. I'm so grateful for your friendship and have learned so much from your many many opinions. A very important thank you to Julia and Audrey; how could any of us have survived this alone!

And of course, thank you to Simone. You've brought me immeasurable joy and comfort. Thank you for never wavering in your support, for always being patient with me, and for all of the adventures. I can't wait to see what's next.

Most importantly, thank you to my family. Without your support, this never would have been possible. Sofia, thank you for always being on my side. Mama and Tato, there are no words to express how grateful I am. You taught me to think for myself and to ask questions, and you always made sure that the world was open before me. For all of my family, these last few years have been hard in many ways, but there's no feeling quite like going home and being surrounded by your love.

Contents

Acronyms

ALICE	A Large Ion Collider Experiment
ATLAS	A Toroidal LHC Apparatus
BSM	Beyond the Standard Model
CB	Combined
CERN	European Center for Nuclear Research
CMS	Compact Muon Solenoid
CP	Charge parity
CR	Control Region
CSC	Cathode-Strip Chamber
CTP	Central Trigger Processor
DV	Displaced Vertex
EM	Electromagnetic
FCAL	Forward Calorimeter
GMSB	Gauge Mediated Supersymmetry Breaking
GSF	Gaussian-sum Filter
HF	Heavy Flavor
HLT	High Level TriggerI
IBL	Insertable B-Layer
ID	Inner Detector
IP	Interaction Point
LAr	Liquid Argon Calorimeter
LEP	Large Electron-Positron Collider
LHC	Large Hadron Collider
LHCb	Large Hadron Collider beauty
LINAC	Linear Accelerator
LLP	Long Lived Particle
LRT	Large Radius Tracking
LSP	Lightest Supersymmetric Particle
L1	Level One
L1Calo	L1 Calorimeter Trigger
L1Muon	L1 Muon Trigger

MC	Monte Carlo simulation
MDT	Monitored Drift Tube
MET	Missing Transverse Energy
MS	Muon Spectrometer
MSSM	Minimal Supersymmetric Standard Model
NLSP	Next Lightest Supersymmetric Particle
OR	Overlap Removal
PDF	Parton Distribution Function
PS	Proton Synchrotron
PSB	Proton Synchrotron Booster
PV	Primary Vertex
QCD	Quantum Chromodynamics
RF	Radiofrequency
RMS	Root mean square
RoI	Region of Interest
RPC	Resistive Plate Chamber
RPV	R-parity violating
SCT	Silicon Microstrip Tracker
SM	Standard Model
SPS	Super Proton Synchrotron
SR	Signal Region
ST	Standard Tracking
SUSY	Supersymmetry
TGC	Thin Gap Chamber
TRT	Transition Radiation Tracker
VR	Validation Region

Part I
Introduction and Motivation

Chapter 1
Introduction

Particle physics seeks to understand the fundamental structure of the universe by defining the minimal set of particles and interactions required to describe all physical phenomena. The Standard Model (**SM**) is the best attempt at such a description. The **SM** has undergone decades of rigorous testing and can explain almost all phenomena we see in experiments. Yet it is known that the **SM** is missing explanations for crucial physical phenomena like quantum description of gravity or a dark matter candidate. As a result, many Beyond the Standard Model (**BSM**) theories have been developed and tested, hoping to extend and complete the picture the **SM** gives. So far, no evidence for any of these theories has been seen.

The Large Hadron Collider (**LHC**) at European Center for Nuclear Research (**CERN**), a 27 km particle collider outside of Geneva, Switzerland, is the largest particle physics experiment in the world, and provides an extremely effective environment to test the **SM** and a wide variety of **BSM** theories. This thesis uses data from the A Toroidal LHC Apparatus (**ATLAS**) experiment, one of the four largest experiments along the **LHC** ring.

Beams of protons circulate and collide in the **LHC**, and if two protons collide with sufficiently high energy, massive particles can be created; $\sqrt{s} = 8$ TeV collisions enabled the 2012 discovery of the Higgs boson with mass of 125 GeV. The **LHC** provides 60 million collisions per second, enabling physicists to search for new and rare physical phenomena. Unfortunately, after 8 years of data taking no evidence of **BSM** physics has been found. Data taking is scheduled to resume in 2022 with only a moderate increase in collision energy and about a factor 2 more data.

This is a call to expand the suite of **BSM** searches by re-examining the assumptions made in searches that have been performed so far. What could we have missed in our search for new physics at the TeV scale? The **LHC** detectors are designed to look the decays of short-lived, heavy particles with the assumption that the decay products will be *prompt* and trace back to the collision point. This misses a large range of intermediate lifetimes of possible **BSM** particles that decay inside

L. Horyn, *A Search for Displaced Leptons in the ATLAS Detector*, Springer Theses,
https://doi.org/10.1007/978-3-030-91672-5_1

of the detector material. These signatures are challenging, but not impossible, to identify with **ATLAS** as they result in *displaced* **SM** particles that do not point back to the collision point. Many **SM** particles are long lived, like muons or neutrons, and many **BSM** theories predict particles with lifetimes that result in displaced decays. This thesis presents a search for one such signature.

This thesis presents a search for two displaced **SM** leptons, either electrons or muons, that are not connected by a displaced vertex [1]. Due to the displacement and lack of vertex, a **BSM** particle decaying to this signature would be vetoed by all other analyses at the **LHC**, even those targetting Long Lived Particles (**LLPs**). This search has unique sensitivity to a specific Gauge Mediated Supersymmetry Breaking (**GMSB**) Supersymmetry (**SUSY**) model where the Lightest Supersymmetric Particle (**LSP**) is the superpartner to the graviton, the gravitino, and the Next Lightest Supersymmetric Particle (**NLSP**) is a slepton ($\tilde{\ell}$), the superpartner to a lepton. The $\tilde{\ell}$ is long lived because it must decay to the gravitino through the very weak gravitational coupling. The last time this model was explored was in the OPAL, ALEPH, DELPHI, and L3 experiments at Large Electron-Positron Collider (**LEP**) [2], where masses up to about 90 GeV were probed for the full range of lifetimes, immediately decaying to detector stable and all possible signatures in between. This search probes almost an order of magnitude of mass phase space in a limited lifetime phase space, and in a significantly more challenging environment than **LEP**.

Since this is the first search for displaced leptons at the **LHC** particle selection algorithms and robust data-driven background estimates needed to be developed. This search for displaced leptons uses 139 fb^{-1} of data collected by **ATLAS** during Run 2 of the **LHC**. Major backgrounds come from fakes of the reconstruction algorithms and muons from cosmic rays. Less than 1 background events are predicted and zero events are seen, and so limits on the mass and lifetime of $\tilde{\ell}$.

This thesis is organized into three main sections: first the search is motivated theoretically, then the experimental setup is described, and finally the search strategy and its results are presented.

Chapter 2 provides theoretical motivation for searches for **SUSY** and in particular long lived and **GMSB SUSY**.

Chapter 3 describes the **LHC** and its design and operation.

Chapter 4 describes the **ATLAS** subdetectors and how they are used to measure particles.

Chapter 5 details particle reconstruction algorithms and the modifications made for this analysis.

Chapter 6 describes the **ATLAS** data and Monte Carlo simulation (**MC**) simulation of the signal model.

Chapter 7 provides context for **LLP** decays and searches.

Chapter 8 provides experimental context for the signature and model.

Chapter 9 describes the analysis strategy, lepton selection requirements and final event selection criteria.

Chapter 10 details the backgrounds to this signature and the algorithms for estimating them.

Chapter 11 describes the uncertainties applied to signal **MC** in order to use the data to make a statement about the **SUSY** model.

Chapter 12 presents the unblinded results.

Chapter 13 discusses the statistical interpretation of the analysis.

Chapter 14 provides reflections on the analysis as well as possible improvements for future **ATLAS** analyses for displaced leptons.

References

1. Search for displaced leptons in $\sqrt{s} = 13$ TeV pp collisions with the ATLAS detector. Technical report. ATLAS-CONF-2020-051. Geneva: CERN (2020). http://cds.cern.ch/record/2740685
2. The OPAL Collaboration, Searches for Gauge-Mediated Supersymmetry Breaking topologies in e^+e^- collisions at centre-of-mass energies up to $\sqrt{s} = 209$ GeV. Eur. Phys. J. C **46**(2), 307–341 (2006). https://doi.org/10.1140/epjc/s2006-02524-8

Chapter 2
Theory

The Standard Model (**SM**) of particle physics provides an extremely successful description of all known particles and interactions. Developed in the 1960s and 70s, the **SM** has survived decades of experimental tests, predicting particles before their discovery and withstanding many high precision tests. However, there are several known shortcomings of the **SM** that lead physicists to believe that nature must deviate from the **SM** at higher energies than currently experimentally accessible.

One attractive extension to the Standard Model (**SM**) is Supersymmetry (**SUSY**), which adds an additional symmetry to the **SM** between fermions and bosons, creating an additional spectrum of particles that provides solutions to the open questions in the **SM**. The particular **SUSY** model probed by this thesis is Gauge Mediated Supersymmetry Breaking (**GMSB**) **SUSY**. This chapter gives an overview of the **SM** and the motivation and formulation of **SUSY**, it then describes **GMSB** models.

2.1 The Standard Model

The Standard Model (**SM**) provides a framework for all known particles and interactions through Quantum Field Theory (QFT). Each particle is represented as a field and the dynamics as a Lagrangian $\mathcal{L}$. All known matter particles and force carriers and the Higgs boson are described through its formulation. The particles of the Standard Model are divided into fermions, made up of leptons and quarks with spin-$\frac{1}{2}$, and bosons with integer spin. Matter is made of fermions and forces are mediated through spin-1 bosons; the spin-0 Higgs boson gives mass to the fermions and other bosons. The particle content of the **SM** is shown in Fig. 2.1.

L. Horyn, *A Search for Displaced Leptons in the ATLAS Detector*, Springer Theses,
https://doi.org/10.1007/978-3-030-91672-5_2

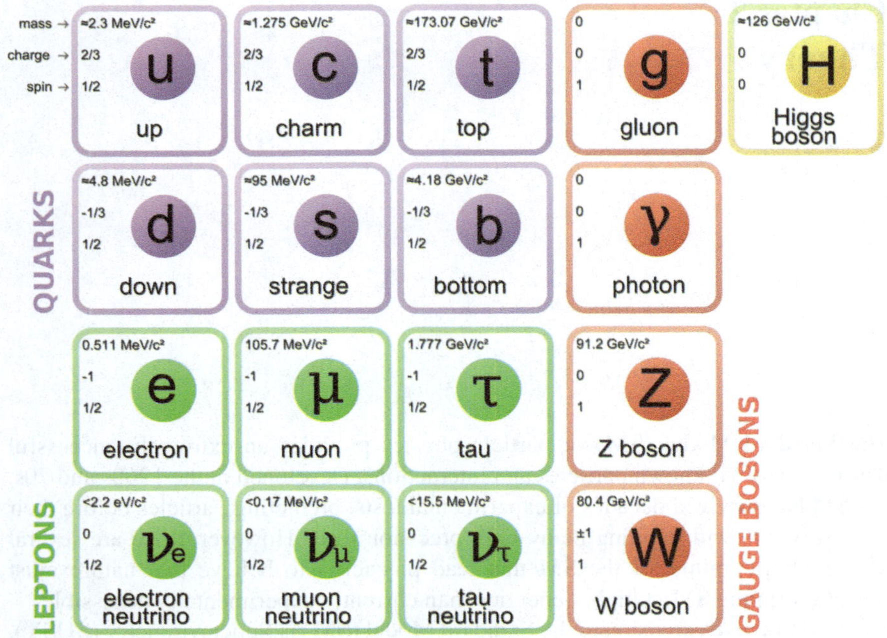

Fig. 2.1 The particles of the Standard Model [1]

2.1.1 Leptons

There are three generations of leptons ℓ, labeled by flavor, which interact via the electromagnetic and weak forces: electrons e, muons μ, and taus τ, ordered by mass. Each lepton has negative electromagnetic charge and has an anti-particle with positive charge.

There are also three generations of leptons ν which interact only via the weak force: electron neutrino ν_e, muon neutrino ν_μ, and tau neutrinos ν_τ, with unknown mass ordering. Neutrinos have no electromagnetic charge and tiny masses and thus are extremely hard to detect and not studied at the **LHC**.

Electrons and neutrinos are stable, while muons and taus are not. Muons decay to an electron and neutrino with a lifetime of 2.2 μs. Taus decay 17% of the time each to an electron or muon and neutrinos, and the remaining 66% of the time into hadrons and neutrinos. In any **SM** process, the number of leptons L of each flavor is conserved.

2.1.2 *Quarks*

There are three generations of quarks that interact via the electromagnetic, weak, and strong forces. Quarks are further divided into up-type with charge $+\frac{2}{3}q_e$ and down-type with charge $-\frac{1}{3}q_e$. The up-type quarks are up u, strange s, and bottom b, the down-type quarks are down d, charm c and top t. Since they also interact via the strong force, each quark also has a color charge, either red, blue, or green (or the opposite—anticolor). Individual quarks are not seen in nature and all stable particles are color neutral with whole number electric charges, either a mix of all three colors or color and anticolor. These color-neutral composite particles are called *hadrons*. Color-anticolor pairs are called *mesons* and states with all three colors are called *baryons*. The lightest hadron, the proton (uud), is stable, but all others are not with a wide range of lifetimes. For example, free neutrons (udd) have a lifetime of about 15 min; pions ($u\bar{u}$) have a lifetime of 26 ns and can travel a substantial distance at the speed of light before decaying; b-mesons a lifetime of 1.5 ps and travel shorter distances if produced in collisions; all other hadrons have shorter lifetimes.

Similar to leptons, the baryon number B is preserved in the **SM**. Baryons have $B = 1$, anti-baryons $B = -1$, and mesons $B = 0$.

2.1.3 *Forces*

The electromagnetic, strong, and weak interactions result from the exchange of bosons between the fermions that make up matter. Force mediators interact with fermions that are charged under the corresponding force's quantum numbers. Gravity is not described by the Standard Model because its relative interaction strength is so small, see Table 2.1 for the relative strength of the fundamental forces.

The symmetries of the Standard Model dictate its interactions. The Standard Model Lagrangian must invariant under choice of gauge. To accomplish this, gauge fields are added to the Lagrangian to cancel out gauge dependences. The gauge fields must be massless or a Higgs field must be introduced in order to maintain the gauge invariance. The Higgs field gives masses to the weak gauge bosons.

The **SM** Lagrangian in invariant symmetries, or local gauge transformations. The symmetry group of the Standard Model is

$$SU_C(3) \times SU_L(2) \times U_Y(1) \tag{2.1}$$

where C stands for the color charge of the strong force, L stands for left because the weak force is left-handed, and Y stands for *hypercharge*, the quantum number of the electroweak force. "Left-handed" describes a particle whose spin is oriented opposite its direction of motion.

Table 2.1 The relative strength of the fundamental forces

Force	Relative coupling strength
Gravity	10^{-39}
Weak	10^{-6}
Electromagnetic	1/137
Strong	1

2.1.3.1 Electroweak Force

The electromagnetic and weak forces are low-energy manifestations of the unified electroweak force. The $SU_L(2) \times U_Y(1)$ symmetry group of the **SM** generates the electroweak force. If the gauge bosons were massless, $SU_L(2)$ alone would describe their interactions. The quantum number of the electroweak force is hypercharge, defined as

$$Y = 2(Q - T_3) \tag{2.2}$$

where Q is the electromagnetic charge and T_3 is the third component of the electroweak isospin, the quantum number of the weak interaction.

The gauge bosons that result from the $SU_L(2) \times U_Y(1)$ symmetry include a triplet and singlet field. The symmetry is broken, these states mix and the electroweak force carriers are produced: $W^{\pm}$ from two of the triplet states, and the Z and photon result from the mixing of the last triplet state with the singlet.

Electromagnetic Force Fermions with electric charge interact via the electromagnetic force mediated by the photon γ. The photon is massless. Being chargeless and spin-1, photons can *pair-produce* a fermion and anti-fermion if the photon's energy is more than twice the rest mass energy of the fermion. Specifically, photons from **LHC** collisions can pair produce electrons and are called *conversions* or *converted photons*.

Weak Force All fermions are charged under the weak force, mediated by the $W^{\pm}$ and Z bosons. The W bosons interact with left-handed particles and right-handed anti-particles, while the Z boson interacts with both left- and right-handed particles. W bosons are charged so they also interact with photons. The W boson has mass of 80.4 GeV and the Z boson of 90.2 GeV.

2.1.3.2 Strong Force

Fermions with color charge interact under the strong force, generated by the $SU_C(3)$ symmetry and mediated by the massless gluon. The gluon itself is color charged, so gluons interact with each other, and the strength of the color force, the strong coupling constant α_s increases as a function of distance. α_s also evolves as a function of energy scale μ as

$$\alpha_s(\mu) = \frac{12\pi}{(32 - 2n_f)\ln\frac{\mu^2}{\Lambda_{\mathrm{QCD}}^2}} \tag{2.3}$$

where n_f is the number of quarks flavors (6) and Λ_{QCD} is the cutoff scale (about 0.2 GeV).

These features of α_s have several important consequences. The first is *color-confinement*. If two quarks are pulled apart, α_s increases and so does the amount of energy required to continue to separate them. Eventually, it becomes energetically favorable to produce a quark-antiquark pair and form two mesons, rather than continue to separate the original quarks. The result of pulling apart the two quarks is not two separated, colored quarks, but two color-neutral hadrons.

Because of this, quarks that result from interactions in the **LHC** cannot be directly observed. If one is produced it first undergoes a *Parton shower*, creating many new colored particles along its trajectory until the particles have enough energy to *hadronize* and form color-neutral particles. This spray of color-neutral hadrons is observed in detectors, and the detector signatures are grouped into a *jet*.

Conversely, at very small distances ($<10^{-16}$ m), the strong coupling is very small, and so quarks inside of hadrons can move around freely. In this regime, Quantum Chromodynamics (**QCD**) can be described perturbatively. Λ_{QCD} is the energy scale at which α_s is too large for perturbation theory to apply.

2.1.4 Higgs Boson

The Higgs boson is the only spin-0 particle in the Standard Model and gives the mass to the gauge bosons via electroweak symmetry breaking and to the fermions via the Yukawa coupling. The addition of the $U(1)$ symmetry group results in a massive gauge field (the electroweak triplet) and a massive scalar (the Higgs). The Higgs can also interact with itself.

Nothing in the **SM** dictates the mass of the Higgs boson. The only known mass scale not yet exploited is the Planck scale, M_P 10^{19} GeV. The mass of the Higgs boson has been measured experimentally at around 125 GeV. The difference between the measured Higgs boson mass and the Planck scale is known as the *hierarchy problem*, which suggests that there is an undiscovered physics scale lower than M_P.

2.1.5 Conserved Quantities

In addition to the gauge symmetry that necessitates the existence of the W and Z bosons, the other symmetries of the Standard Model dictate how its particles interact. Symmetry over translations in time lead to conservation energy, translations in space to conservation of momentum, and rotations in space to conservation of angular momentum. Lorentz invariance leads to the conservation of CPT, the combination conservation of charge, parity, and time. There are other conserved quantities that do not arise from known symmetries of the **SM**, including baryon number B and lepton number L.

2.1.6 Shortcomings

There are many reasons to believe the Standard Model does not give a full picture of fundamental physics, despite its robustness against experimental examination. Perhaps the most obvious hole is the lack of a description of gravity. General Relativity provides an extremely effective description of gravity at large scales, but no such description exists at quantum scales. Gravitational interactions become important at M_P, well beyond the reach of current experiments. Additionally, dark matter and neutrino masses have been observed experimentally, but have no explanation in the Standard Model.

There are also several aesthetic features of the Standard Model that indicate something might be missing, but are not real problems with the theory. The hierarchy problem suggests that there should be another mass scale below the Planck scale to control the Higgs boson mass. The experimentally measured mass is the sum of the bare mass that appears in the Lagrangian and any quantum corrections from interactions with other particles. Should there be no lower mass scale, those quantum corrections would need to be 10^{35} times larger than the Higgs mass. While this is possible, the idea of combining two very large numbers to result in exact small number makes physicists uncomfortable and indicates that there might be something that is not understood, and many *natural* theories that avoid this fine tuning have been proposed.

2.2 Supersymmetry

Supersymmetry (**SUSY**) is a theory proposed and developed in the 1970s [2–4] that provides solutions to many of the problems with the **SM**. **SUSY** adds a symmetry in addition to the ones already discussed: a symmetry between fermions and bosons. An additional dimension is added and rotations in this dimension change the spin of a particle by $\frac{1}{2}$, changing a fermion to a boson and vice versa.

In **SUSY**, each **SM** particle gets a *superpartner* with all of the same quantum numbers except for spin. The bosonic superpartners of fermions are *sfermions* and the fermionic partners of the **SM** bosons have "-ino" appended to their names. For example, the superpartner of a tau τ lepton is a stau $\tilde{\tau}$ and the superpartner of a gluon g is the gluino $\tilde{g}$. A *supermultiplet* refers to the pair of the **SM** particle and its superpartner.

Should **SUSY** exist, we already know that it must be broken. If the selectron had the same mass as the electron it would have been observed decades ago, so superpartners must be heavier than their **SM** counterparts. The mechanism by which **SUSY** can be broken depends on the specific **SUSY** model.

2.2.1 Minimal Supersymmetric Standard Model

The Minimal Supersymmetric Standard Model (**MSSM**) is the simplest self-consistent Supersymmetric extension to the Standard Model. Each **SM** particle gets a **SUSY** partner, except for the Higgs which gets two partners to account for the different masses of the sparticles. One of the Higgs doublets couples to up-type quarks, and the other to down-type quarks and charged leptons.

After electroweak symmetry breaking, there are five Higgs states: A, H, H^+, H^-, h. The lightest state h, is assumed to be the already-discovered **SM** Higgs. The others are called *higgsinos*: A and H are neutral and H^+ and H^- are charged. The higgsinos mix with the electroweakinos (the superpartners of the electroweak gauge bosons). This results in the electrically charged mass-eigenstates called *charginos* and the neutral ones *neutralinos*. The **MSSM** contains four neutralinos ordered by mass $\tilde{\chi}_1^0, \tilde{\chi}_2^0, \tilde{\chi}_3^0, \tilde{\chi}_4^0$ and two charginos ordered by mass $\tilde{\chi}_1^\pm, \tilde{\chi}_1^\pm$.

Each **SM** fermion gets two spin-0 partners called *sleptons*, one right-handed $\tilde{\ell}_R$ and one left-handed $\tilde{\ell}_L$. The left- and right-handed states mix to form mass eigenstates $\tilde{\ell}_1$ and $\tilde{\ell}_2$, ordered by mass. *Squarks*, the super partners of quarks, follow the same pattern as the sleptons, where $\tilde{q}_R$ and $\tilde{q}_L$ mix to form $\tilde{q}_1$ and $\tilde{q}_2$. Each left-handed neutrino has a *sneutrino* superpartner with spin-0 $\tilde{\nu}_L$.

There are many new interactions between particles and sparticles introduced by the **MSSM**. The general rule is that for any **SM** interaction, two of the particles can be replaced by their sparticles. In addition, there are several interactions that are introduced in the **MSSM** that violate the B and L conservation in the **SM**. Specifically, they violate $B - L$ symmetry, which is never violated in the **SM**. To remain consistent with the observed behavior of the **SM**, a new conserved quantity is introduced from a $U(1)$ symmetry: R-parity. It is defined as

$$\text{R-parity} = (-1)^{2s+3B+L} \tag{2.4}$$

where s is spin, B is baryon number, and L is lepton number. All **SM** particles have R-parity $= +1$ and **SUSY** particles have -1. This results in a symmetry of the number of sparticles minus the number of anti-sparticles. Consequently, sparticles must be produced in pairs, and massive sparticles decay through a chain of lighter sparticles and **SM** particles. A given sparticle must always decay to an odd number of sparticles. Some **SUSY** models do not have R-parity as a conserved quantity, though any R-parity violating (**RPV**) coupling must be small in order to maintain consistency with observed physical phenomena. In particular, large **RPV** couplings could lead to proton decay.

2.2.1.1 Relationship to Standard Model Problems

The **MSSM** has 120 free parameters that dictate the masses and interaction strengths of the sparticles. Different assumptions about the **SUSY** breaking mechanism and the allowed couplings change the values of these parameters. The LSP is the lightest Supersymmetric particle in a model, and the NLSP is the next-to-lightest Supersymmetric particle. If R-parity is conserved, the **LSP** is stable because it cannot decay to **SM** particles. If this particle is also electrically neutral, it is an excellent dark matter candidate.

SUSY solves other problems with the **SM** as well. Many physicists believe that the electromagnetic, strong, and weak forces are low energy manifestations of a unified force at high energy, just like the unification of the electromagnetic and weak forces at the GeV scale. When one scales **SM** couplings to high energies, there is no point where all three forces have the same coupling, seen in Fig. 2.2. Performing the same exercise with the **MSSM** with TeV-scale sparticles, the couplings converge. Additionally, the sparticles' interactions with the Higgs cancel out those of the **SM** particles and naturally set the Higgs mass close to its observed value. This relies on the assumption that the mass of the stop quark is similar to that of the top quark to leave a small remaining correction, and requires stop masses at the TeV scale.

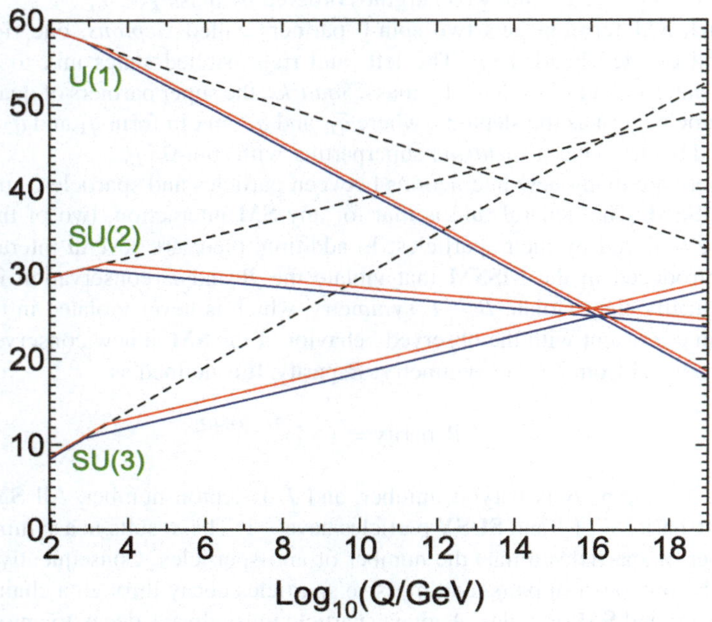

Fig. 2.2 Force coupling strength as a function of energy Q. Their evolution in the **SM** is shown in the dashed lines, and with the **MSSM** in solid lines. The blue lines show sparticle masses around 750 GeV and the red 2.5 TeV. The $SU(3)$ line describes the strong coupling, $SU(2)$ the weak coupling, and $U(1)$ the electromagnetic coupling [5]

These features make **SUSY** and extremely attractive solution to the problems of the **SM**, and TeV-scale particles are perfect targets for searches at the **LHC**. None of the **SUSY** searches performed so far have resulted in any evidence for **SUSY** and have set many limits on **SUSY** parameter space. As more and more searches exclude TeV scale **SUSY** models, ensuring full coverage of possible models is paramount.

2.2.2 *Gauge Mediated Supersymmetry Breaking*

There are various mechanisms by which **SUSY** can be broken. In Gauge Mediated Supersymmetry Breaking (**GMSB**) models, the symmetry breaking occurs via ordinary gauge interactions by heavy messengers carrying $SU(3) \times SU(2) \times U(1)$ quantum numbers [6–8].

Phenomenologically, **GMSB** models have several notable features.

- The **LSP** is always the gravitino $\tilde{G}$ because it does not get its mass through the gauge interactions. $m_{\tilde{G}}$ is proportional to F/M_P, where $\sqrt{F}$ is the **SUSY** breaking scale ($\sim 10^4 - 10^{11}$ GeV) and M_P is the Planck scale ($\sim 10^{19}$ GeV), making the mass extremely small.
- The **NLSP** is either the lightest neutralino $\tilde{\chi}_1^0$ or the slepton $\tilde{\ell}$, depending on the specific model. In the slepton case, the **NLSP** is either the stau $\tilde{\tau}$ or all three flavors $\tilde{e}$, $\tilde{\mu}$, $\tilde{\tau}$ are mass degenerate co-**NLSP**ss. The **NLSP** decay modes are $\tilde{\chi}_1^0 \rightarrow \gamma\tilde{G}$ or $\tilde{\ell} \rightarrow \ell\tilde{G}$. This thesis presents a search for a model with the $\tilde{\ell}$ as the **NLSP**.
- Due to the small gravitational coupling, the decay is suppressed and the **NLSP** can become *long lived*.

This search, like most **SUSY** searches at the **LHC** makes use of a *simplified model*. Full **SUSY** models are incredibly complex and have an enormous number of free parameters (120 in the **MSSM**). Simplified models ignore most particles and parameters, and focus only on a subset of particles and their interactions. In particular, they enable tuning of observable parameters instead of theoretical parameters. The simplified model used in this search describes the interaction shown in Fig. 2.3. The $\tilde{\ell}$ are pair-produced in pp collisions and decay to the $\tilde{G}$ and ℓ. The $\tilde{\ell}$ can be right- or left-handed, and generally the states are assumed to be mass degenerate though the production cross section varies based on the chirality. In the case of a $\tilde{\tau}$, mixing can occur between the right- and left-handed states creating mass ordered $\tilde{\tau}_1$ and $\tilde{\tau}_2$.

The lifetime of the slepton **NLSP** can be expressed as [9]

$$c\tau \approx 100\mu\text{m}\left(\frac{100\text{ GeV}}{m_{\tilde{\ell}}}\right)^5\left(\frac{\sqrt{F}}{100\text{ TeV}}\right)^4 \tag{2.5}$$

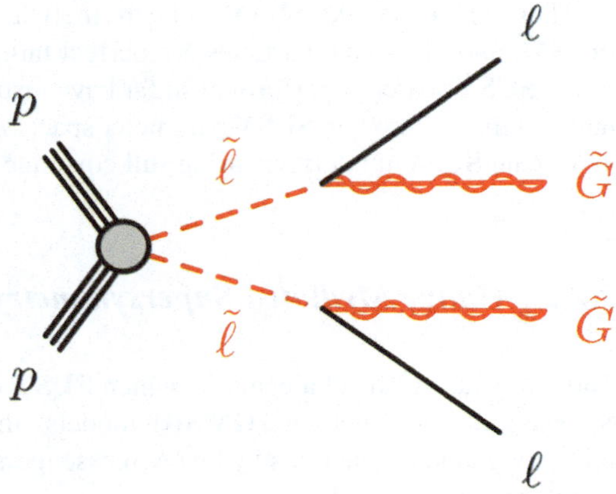

Fig. 2.3 **GMSB** $\tilde{\ell}$ decay to **SM** ℓ and $\tilde{G}$. This is the Feynman diagram targeted by the search presented in this thesis

Even at the low end of possible **SUSY** breaking scales, the $\tilde{\ell}$ will travel a substantial distance relative to **SM** particles [9]. Particle lifetimes will be further discussed in Chap. 7. In the intermediate breaking scale regime, the $\tilde{\ell}$ can travel a small distance before decaying, and the resulting **SM** ℓ is *displaced* meaning it originates a macroscopic distance from the $\tilde{\ell}$ production point. Most **SUSY** searches at colliders assume that leptons are *prompt* and their origin cannot be discerned from the $\tilde{\ell}$ production point and therefore have no sensitivity to $\tilde{\ell}$ with long lifetimes. Long lived $\tilde{\ell}$ production was last explored by the OPAL, ALEPH, DELPHI, and L3 experiments at **LEP** which together excluded models with $\tilde{\ell}$ masses less than about 90 GeV. Experimental details of the LEP searches will be further discussed in Chap. 8.

The search presented in this thesis provides unique sensitivity to **GMSB** $\tilde{\ell}$ decays. The $\tilde{\ell}$ would be pair-produce in an **LHC** collision and the signature of this process is two displaced **SM** ℓ, either two electrons, two muons, or an electron and a muon. The range of possible $\tilde{\ell}$ masses and lifetimes to which this search is sensitive is dictated by:

- The *cross section* of $\tilde{\ell}$ production, or the probability that it will occur in a *pp* collision. The cross section decreases with increasing $\tilde{\ell}$ mass. This signature is not expected to have backgrounds from **SM** processes, so only 3 events must be expected to make an exclusion. This sets an upper bound on the mass sensitivity.
- The size of the detector and reconstruction algorithms dictate lifetime sensitivity. In this search, the trajectory of the lepton must be measured in the tracking detector and that trajectory must be discernibly displaced from the interaction point. This requirement sets the upper and lower bounds of lifetime sensitivity.
- The requirement that both **SM** τ decay leptonically limits the $\tilde{\tau}$ sensitivity relative to $\tilde{e}$ and $\tilde{\mu}$. Due to the branching ratio of the τ, 12% of di-τ events have both τ decaying leptonically.
- The thresholds for triggering an event during data collection set a lower limit on mass sensitivity. These thresholds will be further discussed in Sect. 4.6.

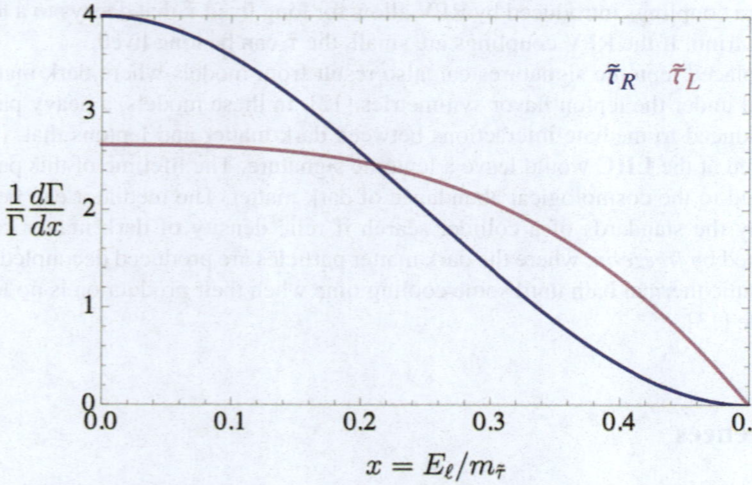

Fig. 2.4 Distribution of τ energies from left- and right-handed $\tilde{\tau}$ decays. A **GMSB** $\tilde{\tau}$ is expected to be right-handed, resulting in lower τ energies [9]

This particularly limits the $\tilde{\tau}$ sensitivity for two reasons. First, electrons and muons from $\tilde{\tau}$ decays have lower momentum than leptons resulting from $\tilde{e}$ or $\tilde{\mu}$ because some of the τ momentum is carried away by the neutrinos in the decay. Additionally, the $\tilde{\tau}$ **NLSP** is expected to be right-handed which results in lower energy τ, shown in Fig. 2.4.

This search is sensitive to $\tilde{e}$ and $\tilde{\mu}$ with masses 50–800 GeV and lifetimes 0.001–10 ns, and $\tilde{\tau}$ with masses 100–400 GeV and lifetimes 0.01–1 ns.

2.2.3 Other Theoretical Models Producing Displaced Leptons

GMSB SUSY is the main theoretical motivation for this search. The displaced dilepton signature provides unique sensitivity to this model in this lifetime regime which is otherwise not covered by any other **LHC** search. However, there are other theoretical **BSM** models which produce displaced leptons [9].

SUSY models with R-parity violating (**RPV**) can result in signatures with displaced leptons [10]. In **RPV** models, the **NLSP** is unstable and can decay into **SM** particles. Since the **RPV** coupling must be small, the **NLSP** can become long lived, and its lifetime is related to the size of the coupling. One such scenario has a stop $\tilde{t}$ **LSP** that decays to a lepton and quark, and simplified models with pair-produced $\tilde{t}$ result in displaced leptons. This signature has been directly in **ATLAS** by Ref. [11] that targeted events with a secondary vertex and a muon. Additionally, the extra couplings introduced by RPV allow for long lived $\tilde{\tau}$ that decays to a lepton and neutrino. If the RPV couplings are small, the $\tilde{\tau}$ can be long lived.

Displaced leptonic signatures can also result from models where dark matter is charged under the lepton flavor symmetries [12]. In these models, a heavy particle is introduced to mediate interactions between dark matter and leptons that, if pair produced at the **LHC** would leave a leptonic signature. The lifetime of this particle is related to the cosmological abundance of dark matter. The mediator can be long lived by the standards of a collider search if relic density of dark matter can be explained by *freeze-in*, where the dark matter particles are produced decoupled from the cosmic thermal bath until some cooling time when their production is no longer possible [13].

References

1. P. Gagnon, The Standard Model: a beautiful but flawed theory (2014). https://www.quantumdiaries.org/2014/03/14/the-standard-model-a-beautiful-but-flawed-theory/
2. J. Wess, B. Zumino, Supergauge transformations in four dimensions. Nucl. Phys. B **70**(1), 39–50 (1974). ISSN: 0550-3213. https://doi.org/10.1016/0550-3213(74)90355-1. http://www.sciencedirect.com/science/article/pii/0550321374903551
3. J.-L. Gervais, B. Sakita, Field theory interpretation of supergauges in dual models. Nucl. Phys. B **34**(2), 632–639 (1971). ISSN: 0550-3213. https://doi.org/10.1016/0550-3213(71)90351-8. http://www.sciencedirect.com/science/article/pii/0550321371903518
4. P. Ramond, Dual theory for free fermions. Phys. Rev. D **3**(10), 2415–2418 (1971). https://doi.org/10.1103/PhysRevD.3.2415
5. S.P. Martin, A supersymmetry primer. Adv. Ser. Dir. High Energy Phys., 1–98 (1998). ISSN: 1793-1339. https://doi.org/10.1142/9789812839657_0001
6. G. Sguazzoni, GMSB susy searches at LEP, in *Proceedings, 35th Rencontres de Moriond on Electroweak Interactions and Unified Theories* (2001), pp. 457–462. 6 p. https://cds.cern.ch/record/2296652
7. S. Dimopoulos et al., Experimental signatures of low energy gauge-mediated supersymmetry breaking. Phys. Rev. Lett. **76**(19), 3494–3497 (1996). ISSN: 1079-7114. https://doi.org/10.1103/PhysRevLett.76.3494
8. S. Ambrosanio, G.D. Kribs, S.P. Martin, Signals for gauge-mediated supersymmetry-breaking models at the CERN LEP 2 collider. Phys. Rev. D **56**(3), 1761–1777 (1997). ISSN: 1089-4918. https://doi.org/10.1103/physrevd.56.1761
9. J. Evans, J. Shelton, Long-lived staus and displaced leptons at the LHC. J. High Energy Phys. **2016** (2016). https://doi.org/10.1007/JHEP04(2016)056
10. R. Barbier et al., R-Parity-violating supersymmetry. Phys. Rep. **420**(1–6), 1–195 (2005). ISSN: 0370-1573. https://doi.org/10.1016/j.physrep.2005.08.006
11. ATLAS Collaboration, Search for long-lived, massive particles in events with a displaced vertex and a muon with large impact parameter in *pp* collisions at $\sqrt{s} = 13$ TeV with the ATLAS detector. Phys. Rev. D **102**(3), 032006 (2020). https://doi.org/10.1103/PhysRevD.102.032006. arXiv: 2003.11956 [hep-ex]
12. P. Agrawal et al., Flavored dark matter, and its implications for direct detection and colliders. Phys. Rev. D **86**(5) (2012). ISSN: 1550-2368. https://doi.org/10.1103/physrevd.86.055002
13. L.J. Hall et al., Freeze-in production of FIMP dark matter. J. High Energy Phys. **2010**(3) (2010). ISSN: 1029-8479. https://doi.org/10.1007/jhep03(2010)080

Part II
The Experiment

Chapter 3
The Large Hadron Collider

The world's largest machine is required to study the universe's smallest particles. The **LHC** is circular particle collider located on the French-Swiss border outside of Geneva, Switzerland. A series of accelerators culminate in a 27-km ring in which beams of hadrons, either protons or heavier ions, are collided. The **LHC** is the most powerful hadron accelerator ever built and has been running in its current form since 2008. It is built in the tunnel previously occupied by the Large Electron-Positron Collider (**LEP**), the most powerful lepton accelerator ever built, which was used to precisely measure the mass of the W and Z bosons. Collider experiments are based around the idea that $E = mc^2$: the more energetic the reaction, the more massive the particle that can be produced by it, allowing scientists to study physical phenomena not accessible in other experiments.

There are four major experiments around the **LHC** ring: the **ATLAS** and Compact Muon Solenoid (**CMS**) experiments, general purpose detectors designed independently in order to serve as a cross-check for each other; A Large Ion Collider Experiment (**ALICE**), a tracking-focused detector designed to study collisions of heavy ions; and Large Hadron Collider beauty (**LHCb**), an asymmetric detector designed to study Charge parity (**CP**) violation.

The dataset used in this analysis comes from proton-proton (pp) collisions during Run 2 of the **LHC**, which spanned from 2015–2018 and had a center of mass energy of $\sqrt{s} = 13$ TeV.

3.1 A Circular Proton-Proton Collider

The history of particle physics is rich with different kinds of accelerators: linear and circular; colliding electrons, positrons, protons, and anti-protons. The **LHC** is a circular, proton-proton collider—a deliberate choice due to its physics goals. These

L. Horyn, *A Search for Displaced Leptons in the ATLAS Detector*, Springer Theses,
https://doi.org/10.1007/978-3-030-91672-5_3

considerations are an important part of the ongoing conversation about how the next generation of particle physics experiments should be designed.

First, accelerators can be linear or circular. In a linear accelerator, a particle is propelled from one end of the beam pipe to the other, increasing its energy as it goes, and eventually collides with a target or beam from another linear accelerator. This means that each particle can only interact with the accelerating field once and then the particles must interact or be dumped. However, in a circular accelerator, particles make many revolutions around the acceleration path, so greater energies can be achieved in the same accelerating distance. In a circular collider, two beams of particles revolve in opposite directions and are focused at specified Interaction Point (**IP**). Particles that do not collide the first time the beams cross can be recycled and be made to collide again, and so collisions can happen continuously for hours. In all, linear accelerators are simpler to create, but are less efficient accelerators than their circular counterparts.

Specifically, **LHC** is a *synchrotron*, composed of many Radiofrequency (**RF**) cavities used to accelerate particles, and many magnets used to focus and bend the beam arranged in a ring. A **RF** cavity is a metallic structure with an electric field that oscillates at a specified frequency. Positively charged particles are repelled through the positive field and pulled by the positive field. Once the particle reaches the required energy, it will not be further accelerated. If the proton has too much or too little energy, it will arrives early or late with respect to the oscillation and it will be decelerated or accelerated by the cavity. This procedure creates a beam made of *bunches* of protons, the size and spacing of which is dictated by the oscillation frequency of the accelerating cavities. At the **LHC** proton bunches are 25 ns apart [1]. Strong electromagnets are used to focus and bend the beam into its circular path. Superconducting,[1] 14 m long, 8.4 T dipole magnets are required to bend the beam into its circular shape. Additional sextupole, octupole and decapole magnets accompany the dipole magnets to correct for edge effects at their extremities. Quadrupole magnets are used to *focus* the beams and squeeze them horizontally and vertically at the collision points [2]. Careful control of the beam is extremely important to avoid radiation damage of the **LHC** and the detectors (Fig. 3.1).

In a synchrotron, the magnetic field magnitude and RF frequency are cycled over time, allowing the particles to remain at a constant radius during acceleration. It uses a series of magnets to bend particles in a ring, unlike its precursors, which use a single magnet and could not reach the same beam energies. Cyclotrons have constant magnetic field magnitudes and RF frequencies and can only be used to accelerate ions due to the effect of relativistic effects on the synchronization between the particle orbit and RF oscillation. Synchrocyclotrons have a cycled RF frequency with a constant magnetic field and could be used to accelerate protons up to about 1 GeV, but higher energies required larger magnets, creating technological and economic limitations [3].

[1] If the magnets were not superconducting, the **LHC** would need to be an order of magnitude larger, 127km, in order to reach the same energy.

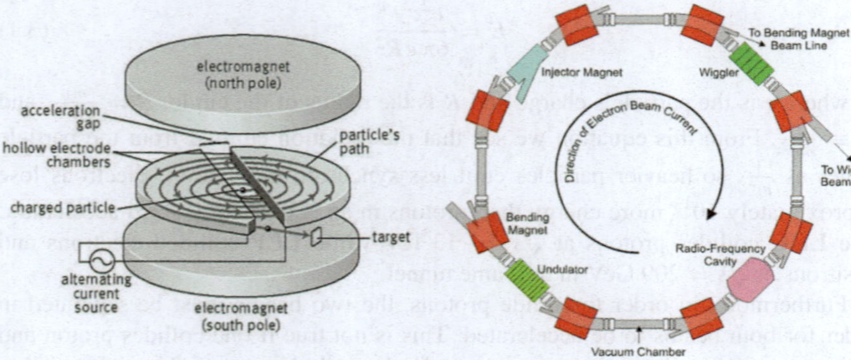

Fig. 3.1 A comparison of schematics for a cyclotron (left) [4] and a synchrotron [5]. The **LHC** is a synchrotron

Secondly, the **LHC** collides protons. Electrons are fundamental particles which create a well understood and clean collision. A collision occurs when an electron and positron annihilate. The energy produced is set by the energy of the beams and there is no ambiguity in the production mechanism of the physical process observed in the detector. Protons, however, are not fundamental particles and when two protons collide, it is their constituent quarks and gluons that interact. This means that not all of the energy of the beam is involved in the primary interaction, nor is the production mechanism known. Additionally, the other quarks and gluons in the proton can interact as well, creating a messier collision environment compared to the electron-positron collision. The fraction of the proton's energy carried by the quarks and gluons is not a known quantity,[2] it must be measured as a Parton Distribution Function (**PDF**), which gives the probability of finding a given constituent with a given momentum. Measuring the **PDF** is difficult and uncertainties in the measurements add to the uncertainties of the final result. The fact that the constituents of protons interact can be seen as an advantage or disadvantage: electron-positron collisions happen at a fixed energy, so they are very powerful for performing precision measurements, for example of the W and Z boson masses, where one would like many clean events where W or Z bosons were produced. In a proton-proton machine, however, a range of energy is available at each collision, making them ideal discovery environments, where one would like to look for a new particle at a wide range of masses, like in the case of the **LHC** searching for the Higgs boson and **BSM** phenomena.

Furthermore, particles moving in a magnetic field lose energy through synchrotron radiation, where the energy lost per turn per particles is given as

[2] In fact, extra *sea quarks* come in and out of existence inside of the protons, these take energy from the main *valence* quarks and can be collide.

$$E = \frac{e^2 c \gamma^4}{6\pi \epsilon R^2} \tag{3.1}$$

where e is the particle's charge and R is the radius of the circle, $R = \frac{mv}{eB\perp}$, and $\gamma \equiv \frac{E}{mc^2}$. From this equation we see that the radiation emitted from the particle scales as $\frac{1}{m^4}$, so heavier particles emit less synchrotron radiation. Electrons lose approximately 10^{13} more energy than protons in an equivalently sized accelerator. The **LHC** collides protons at $\sqrt{s} = 13$ TeV while **LEP** collided electrons and positrons at $\sqrt{s} = 209$ GeV in the same tunnel.

Furthermore, in order to collide protons, the two beams must be separated in order for both beams to be accelerated. This is not true if one collides proton and anti-protons ($p\bar{p}$). Since they are oppositely charged, the same field can be used to accelerate protons in one direction and anti-protons in the other. Proton-anti-proton collisions feature more production mechanisms, but anti-protons are much harder to make and keep than protons. The Tevatron, which discovered the top quark, was a $p\bar{p}$ accelerator.

In principle, one could also make a muon-anti-muon accelerator as well. Muons are heavier than electrons, so do not emit as much synchrotron radiation, and are elementary particles, so provide a cleaner environment than protons and could create TeV-scale collisions. However, muons have a proper lifetime of 2.2 μs and need to be produced in a reactor; to date no such collider has been built.

Additionally, collisions can occur between any stable ion. The **LHC** has collided lead-lead, lead-proton, and xenon-xenon. The additional protons and neutrons make these collisions much more complex, but enables different kinds of physics to be studied. Primarily, it provides high energy and temperature conditions to conditions of the quark-gluon plasma of the early universe, as well as exotic **QCD** states. Additionally, **ATLAS** and **CMS** used the peripheral interactions between lead ions to observe photon-photon scattering [6].

The **LHC** collides protons because they are easy to produce, to keep in the detector for hours, and they produce with a spread of energies across many collisions. It collides them in a circle for efficient, high energy collisions.

3.2 Acceleration

The basic units of an accelerator are **RF** cavities, used to accelerate particles, and magnets, used to focus and bend the beam. A **RF** cavity is a metallic structure with an electric field that oscillates at a specified frequency. Positively charged particles are repelled through the positive field and pulled by the positive field. Once the particle reaches the required energy, it will not be further accelerated. If the proton has too much or too little energy, it will arrives early or late with respect to the oscillation and it will be decelerated or accelerated by the cavity. This procedure creates a beam made of *bunches* of protons, the size and spacing of which is dictated by the oscillation frequency of the accelerating cavities. At the **LHC** proton bunches are 25 ns apart [1].

Strong electromagnets are used to focus and bend the beam into its circular path. Superconducting,[3] 14 m long, 8.4 T dipole magnets are required to bend the beam into its circular shape. Additional sextupole, octupole and decapole magnets accompany the dipole magnets to correct for edge effects at their extremities. Quadrupole magnets are used to *focus* the beams and squeeze them horizontally and vertically at the collision points. The magnet systems are used to cross the beams in the experiments creating *bunch crossings* in which collisions occur. Careful control of the beam is extremely important to avoid radiation damage of the **LHC** and the detectors.

3.3 Injection Chain

A series of accelerators are needed to accelerate protons beams to 6.5 TeV. A schematic of the injection system can be seen in Fig. 3.2.

Protons originate from hydrogen gas that has been passed through an electric field to strip off its protons. They are then injected into the Linear Accelerator (**LINAC**) 2, the only linear accelerator in the chain, that accelerates the protons to 50 MeV, then to the Proton Synchrotron Booster (**PSB**) where they are accelerated to 1.4 GeV, then to 25 GeV in the Proton Synchrotron (**PS**), 450 GeV in the Super

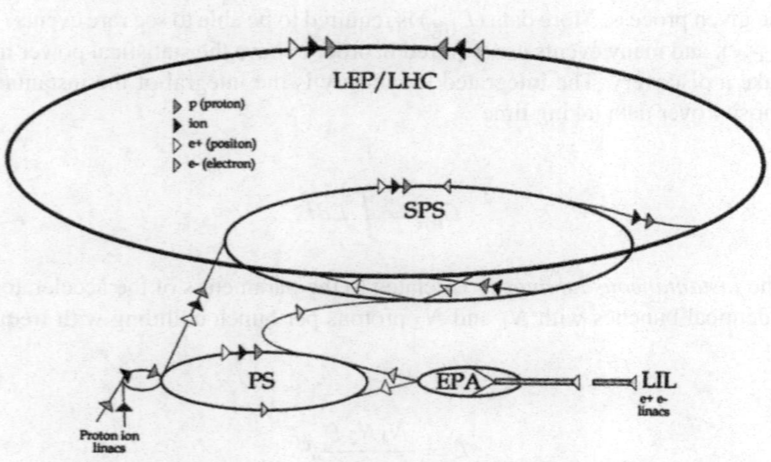

Fig. 3.2 The chain of accelerators used to collide protons at $\sqrt{s} = 13$ TeV in the **ATLAS** detector [7]

[3] If the magnets were not superconducting, the **LHC** would need to be an order of magnitude larger, 127 km, in order to reach the same energy.

Proton Synchrotron (**SPS**). Upon injection to the **LHC**, they are separated into two beams, traveling in opposite directions, where they are finally accelerated 6.5 TeV. Both the **PS** and **SPS** were terminal accelerators at some point in **CERN**'s history. The **PS** was **CERN**'s first synchrotron. The W and Z bosons were discovered in 1983 when the **SPS** was running as a $p\bar{p}$ collider.

It takes about 4 min to fill the **LHC** and 20 min to reach 6.5 TeV, the beam then circulates in the **LHC** and protons collide at the **IPs** for about 8 h. This 8 h period of collisions is called a *run*, while the entire collection of runs taken from 2015–2018 is called *Run 2*.[4] At the end of each run, the beams are dumped, and the process restarts.

3.4 Luminosity

If the primary goal of the collider is to study rare physics processes, whether difficult measurements of the **SM** or heretofore unseen **BSM** physics, enough data must be produced to be able to study them. The number of events of a given process in a given data set is given by

$$N_{\text{events}} = \mathcal{L}_{\text{int}} \times \sigma_{\text{process}} \tag{3.2}$$

where $\mathcal{L}$ is the *integrated luminosity* of the dataset, and σ_{process} is the cross section for the given process. More data ($\mathcal{L}_{\text{int}}$) is required to be able to see rare events (small σ_{process}), and many events are required in order to have the statistical power in able to make a discovery. The integrated luminosity is the integral of the instantaneous luminosity over data taking time

$$\mathcal{L}_{\text{int}} = \int \mathcal{L} dt \tag{3.3}$$

and the *instantaneous luminosity* is related to the parameters of the accelerator. For two identical bunches with N_1 and N_2 protons per bunch colliding with frequency f:

$$\mathcal{L} = \frac{N_1 N_2 f}{4\pi \sigma_x^* \sigma_y^*} \mathcal{F} \tag{3.4}$$

[4] *Run 1* took place from 2009–2013 with a center of mass energy of 7 TeV and resulted in 27 fb^{-1} of data. This is the run in which the Higgs boson was discovered. Run 1 data is not used in this analysis.

Table 3.1 Beam parameters for 2018 for standard running conditions used in the data collection for this analysis. Special runs take place where these parameters are changed

Parameter	Value
N_b (protons per bunch)	1.1×10^{11}
n_b (bunches per beam)	2556
β^* (beam size)	.3 m
ϵ_n (beam spread)	1.8–2.2 μm-radians
$\mathcal{L}_{\text{peak}}$ (peak instantaneous luminosity)	$21 \times 10^{33}\,\text{cm}^{-2}\text{s}^{-1}$
Space between bunches	25 ns

where σ_x^* and σ_y^* are the Root mean square (**RMS**) of the beam width in the x and y directions, and $\mathcal{F}$ is a factor that takes in other geometric effects such as the crossing angle and bunch length, it is generally $\mathcal{O}(1)$. It can be rewritten more specifically for the **LHC** as

$$\mathcal{L} = \frac{N_b^2 n_b f}{4\pi\sqrt{\epsilon_n \beta_x^* \beta_y^*}}\mathcal{F} \tag{3.5}$$

where N_b is the number of protons per bunch (assuming $N_1 = N_2$), n_b is the number of bunches in the beam, ϵ_n is the *emittance*, which describes the spread of the particles in the bunch, and β^* is the value of the β-function at the **IP**. The β-function describes the size of the beam as a function of location. A sampling of the values of these parameters in 2018 are shown in Table 3.1 [8].

Generally, the luminosity decreases over the course of the run as protons are collided in the bunches. but several of these factors can be manipulated *in situ*, increasing or decreasing the luminosity depending on the need. Decreasing β^* increases the luminosity, so the beams are *squeezed* with focusing quadrupole magnets at the **IP**. Nonzero beam crossing angles and longer bunches decrease luminosity. During some runs of Run 2 of the **LHC**, the beam angle was changed at the beginning of the run in order to decrease the luminosity to a level more tolerable by the experiments, referred to as *leveling*.

In **ATLAS**, luminosity is measured in two steps. First, the rate of pp collisions (μ_{vis}) is measured using detectors close to the beam pipe. This is done both online, so that adjustments to the data collection scheme can be done on the fly, as well as offline, to determine the amount of data collected. Second, the pp collision rate is translated into a luminosity using a *van der Meer* scan. During the scan, the beam is widened and, initially, the beams are separated in both x and y. The beams are moved incrementally closer to each other by known amounts such that the separation between the beams is always known as the number of collisions increases. ATLAS Collaboration [9] The luminosity per bunch can be expressed as

$$\mathcal{L}_b = \frac{\mu_{\text{vis}}}{\sigma_{\text{vis}}} f \tag{3.6}$$

where μ_{vis} is measured by the luminosity detectors. The luminosity, $\mathcal{L}_b$, is known during the van der Meer scan, and so the inelastic cross section, σ_{vis} can be determined and used to calculate $\mathcal{L}_b$ from μ_{vis}. Both μ_{vis} and σ_{vis} contain detection efficiency effects [10].

In Run 2, the **LHC** produced an integrated luminosity of 156 fb^{-1}. **ATLAS** does not operate with perfect efficiency, so 147 fb^{-1} was recorded. Ultimately, 139 fb^{-1} could be for physics analyses after final data quality filtering.

3.4.1 Pileup

While high instantaneous luminosity enables the fast accumulation of data, it also creates a dense environment in which those processes occur. The number of concurrent *pp* collisions is called *pileup*; this number is substantial because the instantaneous luminosity is greater than the *pp* inelastic scattering cross section. The average number of interactions per bunch crossing is not a static number, but the average is 34 for all of Run 2, as seen in Fig. 3.3.

In practice, pileup collisions are generally lower energy, **QCD**-only interactions, producing sprays of low energy activity in the Inner Detector (**ID**) and calorimeters, which can increase the probability of creating fake tracks, clusters, or add energy to non-pileup detector signatures. Pileup can be mitigated by reconstructing all of the vertices in the event and removing detector signatures associated to non-primary vertices. An example of an event with 25 simultaneous collisions can be see in

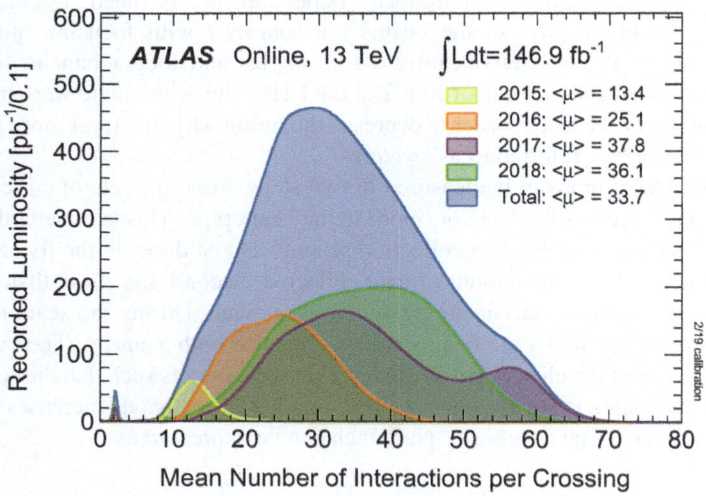

Fig. 3.3 Average number of interactions per bunch crossing during Run 2 [11]

Fig. 3.4 An event display with 25 simultaneous interactions. The primary interaction is in yellow, but the pileup-dense environment makes it hard to see that event among all of the other activity [12]

Fig. 3.4. The tracks from the target interaction are the bold yellow lines, however there are many other tracks to confuse the situation.

References

1. CERN, Accelerating: radiofrequency cavities (2020). https://home.cern/science/engineering/accelerating-radiofrequency-cavities
2. L. Rossi, Superconducting Magnets, European Organization for Nuclear Research European Laboratory for Particle Physics The LHC Superconducting Magnets (2003)
3. Synchrotrons and Cyclotrons (2005). http://www.geology.wisc.edu/~johnf/g777/Misc/chap15.pdf
4. A schematic diagram of a Cyclotron (2013). https://www.researchgate.net/figure/Fig-19-A-schematic-diagram-of-a-Cyclotron_fig4_280722233
5. How do Synchrotrons Work? (2006). http://pd.chem.ucl.ac.uk/pdnn/inst2/work.htm
6. The ATLAS Collaboration, Evidence for light-by-light scattering in heavy-ion collisions with the ATLAS detector at the LHC. Nat. Phys. **13**(9), 852–858 (2017)
7. P. Vanlaer, Contribution to the study of the central tracking system of the CMS detector, at the future proton collider LHC (2020)
8. M. Tanabashi et al., Review of particle physics. Phys. Rev. D **98**(3), 030001 (2018). https://doi.org/10.1103/PhysRevD.98.030001
9. ATLAS Collaboration, Luminosity determination in *pp* collisions at $\sqrt{s} = 7\,\text{TeV}$ using the ATLAS detector at the LHC. Eur. Phys. J. C **71**, 1630 (2011). https://doi.org/10.1140/epjc/s10052-011-1630-5. arXiv: 1101.2185 [hep-ex]
10. ATLAS Collaboration, Luminosity Determination Using the ATLAS Detector. ATLAS-CONF-2010-060 (2010). https://cds.cern.ch/record/1281333
11. ATLAS Experiment, LuminosityPublicResultsRun2 (2012). https://twiki.cern.ch/twiki/bin/view/AtlasPublic/LuminosityPublicResultsRun2
12. ATLAS Experiment, ATLAS Stand-Alone Event Displays (2012). https://twiki.cern.ch/twiki/bin/view/AtlasPublic/EventDisplayStandAlone

Chapter 4
The ATLAS Detector

The ATLAS detector is a cylindrical general purpose particle detector designed to measure the products of $\sqrt{s} = 14$ TeV proton-proton collisions at the **LHC**. It consists of three major sub-detectors: closest to the beamline is the Inner Detector (ID), which measures the trajectories of charge particles, followed by the Calorimeters, which measure the energies of electromagnetic and hadronically interacting particles, and finally the Muon Spectrometer (MS) which measures the trajectories of muons. The **ID** is surrounded by a super conducting solenoid that provides a uniform 2 T magnetic field, enabling measurement of particles' charge and momentum. A toroidal magnet surrounds Muon Spectrometer (**MS**), for charge and momentum measurements of muons. In general, each subdetector consists of a barrel detector parallel to the beamline and perpendicular end-cap detectors on either end of the barrel detectors [1].

A schematic of the **ATLAS** detector is shown in Fig. 4.1.

4.1 Coordinate System

ATLAS uses a Cartesian right-handed coordinate system. The z-axis points along the beampipe, where $+z$ points counter-clockwise. The transverse plane, the y-axis and x-axis, points upward and toward the center of the **LHC** ring, resspectively. The detector is built with symmetry across the origin in in z, as well as with rotational symmetry in the transverse plane. The origin is the pp collision point, generally somewhere near $x = y = z = 0$. The $+z$ side of the detector is referred to as the A-side, and $-z$ as the C-side.

Cylindrical coordinates are generally used to describe the **ATLAS** detector, where ϕ measures the angle in the $x - y$ plane around the beam axis, and θ the angle from the z axis. ϕ is positive for positive y.

L. Horyn, *A Search for Displaced Leptons in the ATLAS Detector*, Springer Theses,
https://doi.org/10.1007/978-3-030-91672-5_4

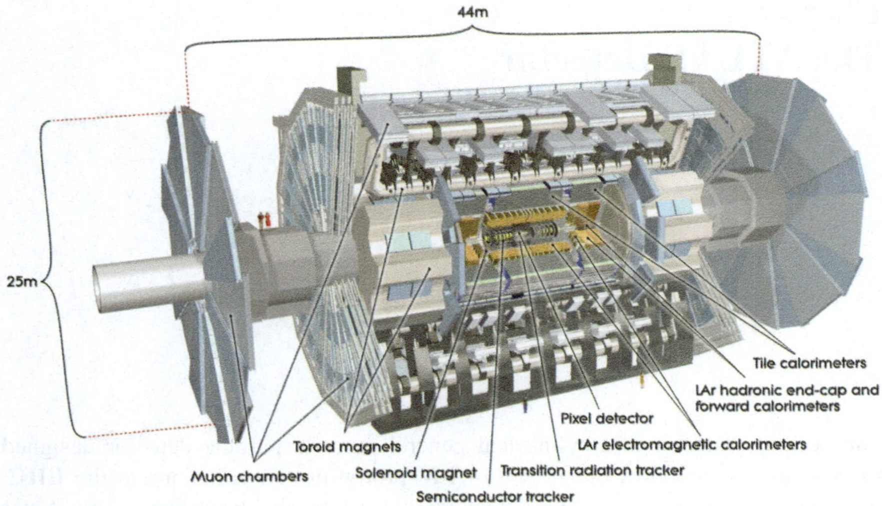

Fig. 4.1 A diagram of the **ATLAS** detector. The dimensions, subdetectors, and magnet systems are labeled [2]

A given particle's momentum in the z direction is not known, but since the beams collide head on along the z-axis, its transverse momentum must be 0, so it is advantageous to define spatial variables independent of z momentum. Thus, instead of θ, *pseudorapidity* is used to describe angle from the z axis. η is defined as

$$\eta = -\ln(\tan\frac{\theta}{2}) \tag{4.1}$$

Particles perpendicular to the z axis have $\eta = 0$, while those parallel to the beamline have $\eta \to \infty$. The particle distribution from **LHC** collisions is roughly uniform in η and η is invariant under Lorentz boosts assuming massless particles.

Angular distances between objects is described using $\Delta R = \sqrt{\Delta\eta^2 + \Delta\phi^2}$ and the radial distance from the origin in the $x - y$ plane is denoted R.

A particle's momentum will generally be described in terms of its p_{T}, its momentum in the transverse direction. A particle's 3-vector is described by $(p_{\mathrm{T}}, \eta, \phi)$, which are all invariant under boosts in z assuming the particle can be considered massless.

4.2 Inner Detector

The Inner Detector measures the trajectories of charged particles resulting from **LHC** collisions. The **ID** covers the region with $|\eta| < 2.5$, measuring approximately 1000 particles per bunch crossing. In order to achieve the momentum and vertex resolution required to achieve **ATLAS**'s physics goals three subdetectors are used: the Pixel detector, the Silicon Microstrip Tracker (**SCT**), and the Transition Radiation Tracker (**TRT**). The Pixel and **SCT** detectors are used for high granularity precision tracking and the **TRT** is used to distinguish electrons from low-mass hadrons like pions. All of this is immersed in a 2T magnetic field, bending the trajectories of charged particles in proportion to their momentum in the ϕ direction.

The p_T resolution of the **ID** scales with track p_T. Higher p_T tracks are less curved, so the measurement resolution is worse. In the **ATLAS ID**, the p_T resolution $0.05\% \times p_T$ with a 1% constant term. The constant term describes measurement uncertainties that do not scale with momentum or energy, such as material imperfections, non-uniform detector response, or other constant measurement issues and is added in quadrature ($\oplus$) with the stochastic term.

The combination of the three trackers with radii ranging from 32.33 mm to 1082 mm enable robust and precise pattern recognition of all charged particles that pass through the detector. The Pixel and **SCT** detectors provide secondary vertexing and impact parameter measurement and vertexing for heavy-flavor jet and τ-lepton tagging. The **TRT** has lower precision per point, but measures a longer track and contributes significantly to momentum measurements.

A sketch of the **ID** can be seen in Fig. 4.2 and a detailed schematic of the various subdetectors is shown in Fig. 4.3.

The two innermost detectors, the Pixel and **SCT** detectors, are made of silicon, which is the preferred tracking medium of particle physics. The **ATLAS** pixels are n-type semiconductors, which have added impurities to increase the number of potential electron charge carriers. A voltage is applied, 150–600 V in the Pixel

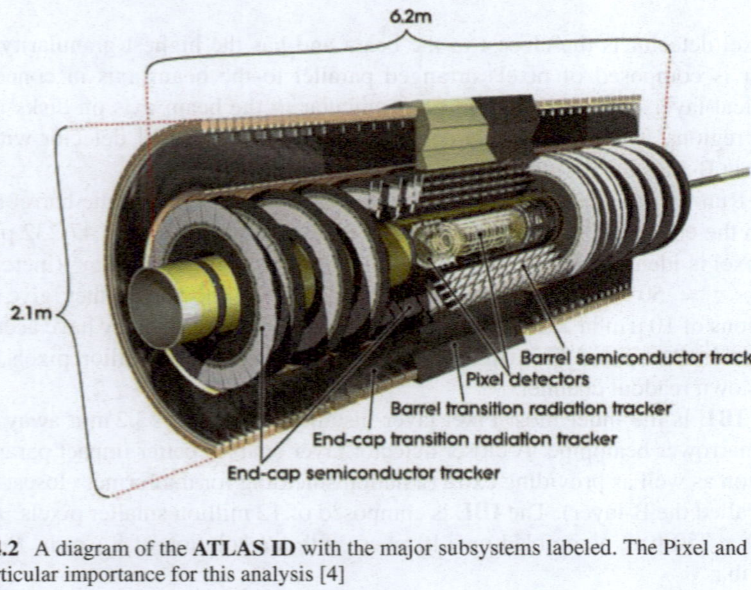

Fig. 4.2 A diagram of the **ATLAS ID** with the major subsystems labeled. The Pixel and **SCT** are of particular importance for this analysis [4]

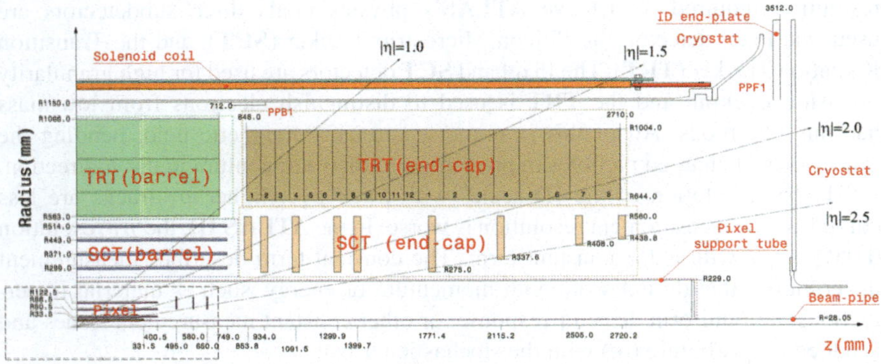

Fig. 4.3 A schematic of the **ATLAS ID** shown in the $R - z$ plane [5]

detector. When a charged particle interacts with a silicon detector layer, electron-hole pairs are created and these pairs drift toward the readout chips in the magnetic field. If enough of these pairs are created such that the signal becomes greater than a pre-specified noise threshold, the signal is readout as a *hit*. A particle must have 3.6 eV to induce about 80 electron-hole pairs per μm. Pixels provide a high signal-to-noise ratio, and so are preferred in regions where high occupancy is possible, like the region closest to the beamline [3].

4.2.1 The Pixel Detector

The Pixel detector is the closest to the beam and has the highest granularity. The detector is composed of pixels arranged parallel to the beam axis in concentric cylindrical layers in the barrel and perpendicular to the beam axis on disks in the endcap regions. The current Pixel detector consists of the Run 1 detector with the Insertable B-Layer (**IBL**) installed closer to the beam pipe.

The Run 1 Pixel detector has 3 concentric cylindrical layers in the barrel and 3 disks in the endcap. It is composed of 1744 pixel sensors, each with 47,232 pixels. Each pixel is identical has minimum size of $R\phi \times z = 50 \times 400\ \mu\text{m}^2$ (increasing to $R\phi \times z = 50 \times 600\ \mu\text{m}^2$ near a readout chip). In the barrel, they give track resolutions of 10 μm in $R\phi$ and 115 μm in z, and in the endcaps they have accuracy of 10 μm in $R - \phi$ and 115 μm in R and the detector has 80.4 million pixels, each with its own readout channel.

The **IBL** is the inner most Pixel layer installed for Run 2 33.2 mm away from a new, narrower beampipe. A closer detector layer enables better impact parameter resolution as well as providing extra radiation shielding for the former closest Pixel layer (called the B-layer). The **IBL** is composed of 12 million smaller pixels, $R\phi \times z = 50 \times 250\ \mu\text{m}$, assembled on 14 staves, with a resolution of 8 μm in $R\phi$ and 40 μm in z.

4.2.2 *The Silicon Microstrip Tracker*

A particle moving through the **SCT** crosses 4 bilayers of the detector creating 4 hit measurements as it interacts with 8 layers of material. A bilayer is composed of 2 stereostrips composed of two thin layers of silicon: one parallel to the beam axis (measuring $R\phi$) and the other rotated by 40 mrad. The combination of the stereostrips in the bilayer gives a three-dimensional measurement. In the barrel, they are arranged in concentric cylinders parallel to the z axis and in the end caps, they are arranged on 9 disks per side. The strips are 12 cm long and 80 µm thick.

Less precise than the Pixels, they have accuracy of 17 µm in $R-\phi$ and 580 µm in z in the barrel region, and in the endcaps they have resolution of 17 µm in $R-\phi$ and 580 µm in R. The use of strips instead of pixels reduces the number of readout channels as well as the cost: there are 15,192 sensors each with 768 strips with an applied voltage of 150–300 V and 6.3 million readout channels.

4.2.3 *The Transition Radiation Tracker*

The **TRT** is straw-tube tracker that extends the tracking volume by almost 500 mm without the cost of that much additional silicon. The tubes have a diameter of 4 mm and made of Kapton and carbon fibers. They are filled with a gas mixture (70% Xe, 27% CO_2, and 3% O_2) and a 31 µm diameter gold-plated tungsten wire. The wires are divided into two halves about $\eta = 0$. In the barrel, the 52,544 straws are parallel to the beam axis and 144 cm long. In the endcaps, 122,880 37 cm long straws are arranged radially in wheels. The **TRT** 351,000 readout channels, much fewer than the silicon detectors.

When a charged particle passes through the tube, the gas is ionized and the resulting free electrons drift toward the wire where they are amplified then readout. A particle passing through the **TRT** typically leaves 36 hits. The **TRT** only provides $R\phi$ measurements with an accuracy of 130 µm per straw. However, it has the advantage that its radius improves the momentum resolution, and it can provide ns-level timing information.

4.2.4 *Solenoid Magnet*

The central solenoid surrounds the **ID** and provides a uniform 2 T field that bends the trajectories of charged particles. The transverse momentum of the particle can be inferred from its radius of curvature, R, in the $x-y$ plane using the equation $p_T = qBR$, where q is the charge of the particle and B the magnetic field in the z direction.

However, the placement of the solenoid between the **ID** and calorimeters necessitates careful design choices so that all of a given particle's energy is still measured by the calorimeters. The solenoid only contributes about 0.66 *radiation lengths*, the mean distance in which an electron loses all but $\frac{1}{e}$ of its energy. In order to achieve this, the solenoid and Electromagnetic (**EM**) calorimeter share a vacuum vessel, eliminating the need for two vacuum walls. It is made of Al-stabilised NbTi superconductor which allows a high electric field to be achieved (7.730 kA) while optimizing the thickness of the coil. The solenoid has an axial length of 5.8 m and radial thickness of 100 cm and it operates at a temperature of 4.5 K.

4.3 Calorimeters

The **ATLAS** calorimeters measure the energy of particles which interact via electromagnetic and hadronic interactions. The calorimeter system is composed of **EM**, hadronic, and forward calorimeters. While they use a variety of different technologies to measure energies, they are all sampling calorimeters composed of alternating active and absorbing layers. Particles shower in absorbing layers and the showers are measured in the active layers. The actual energy of each particle is not measured due to energy loss in the absorbing layers. The size of each calorimeter is set by its radiation length or nuclear interaction length such that the calorimeter absorbs all of given particle's energy by the time it reaches the end of the calorimeter. The only Standard Model particles that should escape the calorimeters are neutrinos and muons; neutrinos do not interact with the material of any of the detectors, and muons are too heavy to be stopped by the EM calorimeter. Unlike the tracker, the energy resolution of a calorimeter increases with increasing energy due to the increased signal generated. A schematic of the **ATLAS** calorimeters can be seen in Fig. 4.4.

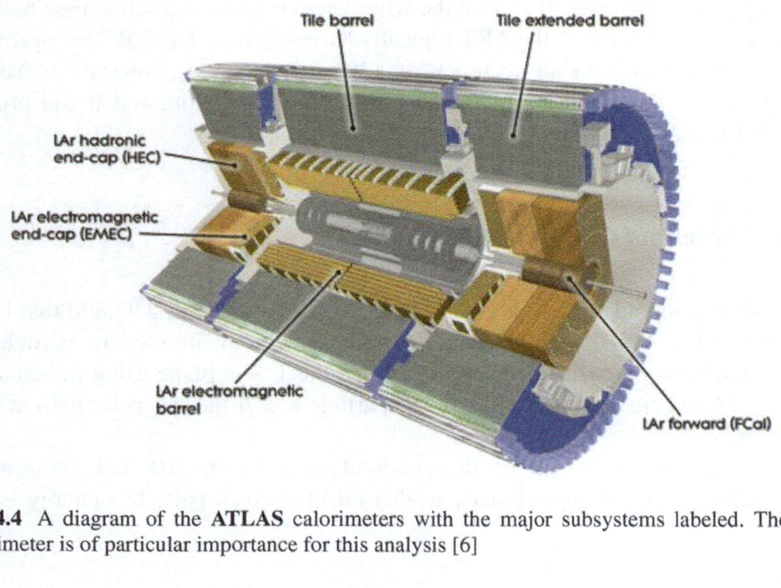

Fig. 4.4 A diagram of the **ATLAS** calorimeters with the major subsystems labeled. The **EM** calorimeter is of particular importance for this analysis [6]

4.3.1 Electromagnetic Calorimeter

The Liquid Argon (LAr) Electromagnetic (EM) calorimeter is the innermost calorimeter and gives excellent energy and position resolution [6]. It is composed of barrel ($|\eta| < 1.5$) and end-cap ($1.4 < |\eta| < 3.2$) components. Both components use a lead absorber with liquid argon active material. The layers of the calorimeter have an accordion shape (shown in Fig. 4.5), which allows for multiple absorbing layers without any gaps between them, as well as complete ϕ symmetry. The first layer has finer segmentation in η to allow for more precise angular measurements of photons (which do not produce an **ID** track). The thickness of the absorbing plates varies as a function of η to optimize energy resolution. A liquid argon active presampler is placed before the accordion layers in the region with $|\eta| < 1.8$ to correct for energy loss upstream of the calorimeter. The whole calorimeter is about 22 radiation lengths wide and gives energy resolution of $10\%/\sqrt{E} \oplus 0.7\%$

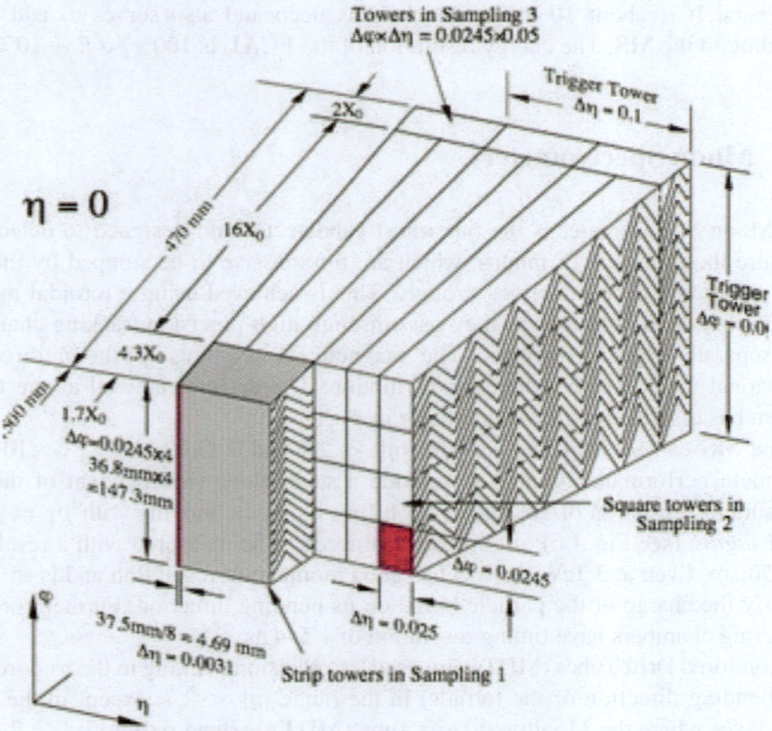

Fig. 4.5 A diagram of the **ATLAS EM** Liquid Argon Calorimeter (**LAr**) calorimeter. It has a unique shape in order to provide precision position and energy resolution [7]

4.3.2 Hadronic Calorimeter

The hadronic calorimeter surrounds the **EM** calorimeter also within $|\eta| < 3.2$. The barrel region ($|\eta| < 1.7$) is made of steel absorbers with active material of scintillating tiles. Here, the calorimeter is about 2 m long in the radial direction and covers about 8 interaction lengths. The Hadronic End-cap Calorimeters cover the region $1.5 < |\eta| < 3.2$ with a copper absorber and liquid argon active material.

The hadronic calorimeter has an energy resolution of $50\%/\sqrt{E} \oplus 3\%$.

4.3.3 Forward Calorimeter

The Forward Calorimeter (**FCAL**) is measures both **EM** and hadronic energy and extends coverage to $3.1 < |\eta| < 4.9$. The detector uses liquid argon as its active material with copper (for **EM** activity) and tungsten (for hadronic activity) absorbers. It is about 10 interaction lengths deep and also serves to add some shielding to the **MS**. The energy resolution of the **FCAL** is $100\%/\sqrt{E} \oplus 10\%$.

4.4 Muon Spectrometer

The Muon Spectrometer is the outermost subdetector and designed to detect and measure the momenta of muons, which are too massive to be stopped by the EM calorimeter and do not interact strongly. This is achieved using a toroidal magnet system that deflects muons as they pass through high-precision tracking chambers and separate trigger chambers. The magnetic field points in the η direction, orthogonal to the muon's direction of motion. The detector as well as the toroid system has discrete rotational symmetry in ϕ.

The **MS** can measure muons with $|\eta| < 2.4$ and 3 GeV $< p_{\mathrm{T}\mu} <$ 10 TeV. The main performance goal is to provide a stand-alone (independent of the **ID**) momentum resolution of 10% at $p_\mathrm{T} >$ 1 TeV; a particle moving with $p_\mathrm{T} =$ 1 TeV has a *sagitta* (see Fig. 4.6) of 500 μm that needs to be measured with a resolution of <50μm. Even at 3 TeV, the **MS** has good momentum resolution and is still able identify the charge of the particle based on its bending direction. Furthermore, the triggering chambers have timing resolution of 1.5–4 ns.

Monitored Drift Tubes (MDTs) are used for precision tracking in the η coordinate (the bending direction of the toroids) in the range $|\eta| < 2.7$, except in the inner most layer where the Monitored Drift Tube (**MDT**)s extend only to $|\eta| < 2.0$ and Cathode-Strip Chambers (CSCs) cover the region $2.0 < |\eta| < 2.7$. Triggering and ϕ measurements are provided by Resistive Plate Chambers (RPCs) in the range $|\eta| < 1.05$ and Thin Gap Chambers (TGCs) in $1.05|\eta| < 2.7$. The detector chambers are arranged in concentric cylindrical shells around the beam axis at radii of about

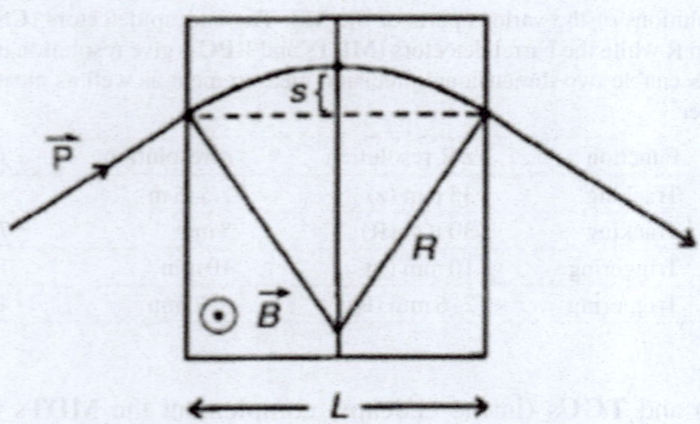

Fig. 4.6 The trajectory of a particle in a magnetic field ("B"). The particle's sagitta, labeled "s" is the bending distance in the direction perpendicular to the magnetic field. The direction of the bending is dictated by the charge of the particle. The higher the momentum of a particle, the smaller the sagitta [8]

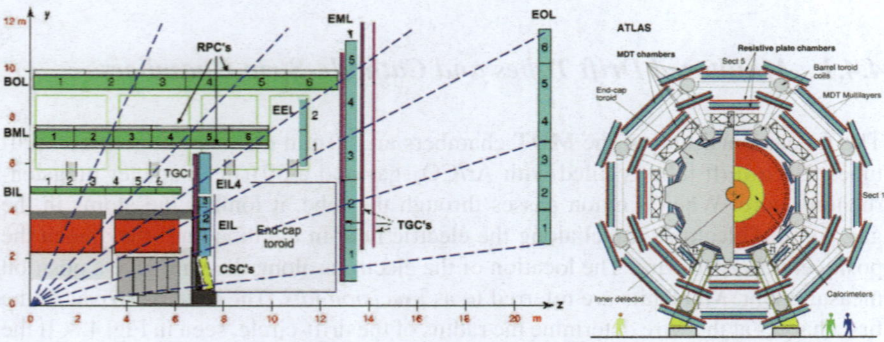

Fig. 4.7 A diagram of the **ATLAS MS** in the $r - z$ (left) plane and a ϕ cross section of the barrel region. On the left, the barrel **MDT** are shown in green and the Resistive Plate Chamber (**RPC**) shown in black. In the end-caps, the **MDT** are shown in blue and the Thin Gap Chamber (**TGC**) shown in purple. On the right, the concentric cylindrical layers of **MDT**s are shown: 8 small chambers and 8 large chambers. The diameter is 20 m at its widest [9]

5 m, 7.5 m and 10 m in the barrel, and on wheels perpendicular to the beam axis at distances of about 7.4 m, 10.8 m, 14 m, and 21.5 m from the interaction point. There is a gap in detector coverage around $|\eta| < 0.1$ for detector access for service work. Schematics of the **MS** can be seen in Fig. 4.7.

While the **MDT**s give extremely precise measurements of muon directions, they cannot be used in isolation for the physics goals of **ATLAS**. First, they are finely segmented in η, the bending direction of the muons, enabling the high momentum resolution. However, in ϕ, their resolution is as good as the length of the tube, several meters long or $\approx$0.1 in ϕ. Additionally, being drift tubes, their readout is very slow and cannot be used for triggering, which requires a quick response. Thus, **RPC**s

Table 4.1 Resolutions of the various parts of the **MS**. The endcap detectors (**CSC**s and **TGC**s) give resolution in R while the barrel detectors (**MDT**s and **RPC**s) give resolution in z. The various pieces of the **MS** enable two dimensional precision measurement as well as measurements to be used in the trigger

Detector	Function	z/R resolution	ϕ resolution	Time resolution
MDT	Tracking	35 μm (z)	2.5–5 m	<700 ns
CSC	Tracking	30 μm (R)	5 mm	7 ns
RPC	Triggering	10 mm (z)	10 mm	1.5 ns
TGC	Triggering	2–6 mm (R)	3–7 mm	4 ns

(in the barrel) and **TGC**s (in the endcaps) complement the **MDT**s for triggering and Cathode-Strip Chamber (**CSC**)s complement the **MDT**s in the forward region (where the particle flux is the highest) to provide better two-dimensional position measurements. A comparison of the resolution of the different detectors of the **MS** can be seen in Table 4.1.

4.4.1 Monitored Drift Tubes and Cathode-Strip Chambers

The fundamental unit of the **MDT** chambers are 30 mm diameter pressurized drift tubes. Each drift tube is filled with Ar/CO_2 gas and a 50 μm diameter tungsten-rhenium wire. When a muon passes through the tube, it ionizes the atoms in the gas. Ionized electrons travel along the electric field in the tube until they reach the positively charged wire. The location of the electrons along the wire give a position measurement. **MDT** hits are referred to as *precision hits*. The time of arrival of the first charges at the wire determine the radius of the drift-circle, seen in Fig. 4.8. If the muon itself passes close to the wire during its trajectory it can induce a signal in the wire, so a dead time is specified before the detector readout. The spatial resolution of a single drift tube is 80 μm.

Each **MDT** chamber consists of two multilayers of drift tubes separated by a mechanical spacer. The inner chambers, where the highest performance is required, a multilayer is composed of four layers of drift tubes (increasing the resolution to 30 μm), while in the middle and outer chambers multilayers are composed of three layers (increasing the resolution to 35 μm). The combination of the three chambers gives the required sagitta resolution with about 15 measurements per muon. There are 1150 **MDT** chambers with 354,000 drift tubes. To achieve the required momentum resolution, the tubes must not bend more than 100 μm, a significant challenge against gravity. The tubes are held in tension and carefully calibrated and monitored to achieve a maximum bending of 20 μm.

The **CSC**s are multiwire proportional chambers with anode wires oriented radially. The wires are held between two cathodes: one segmented with strips perpendicular to the wires giving a precision measurement in the bending plane (η) of with resolution of 40 μm and the other segmented parallel to the wire, giving

Fig. 4.8 A sketch of a muon passing through a cross section of an **MDT** [1]

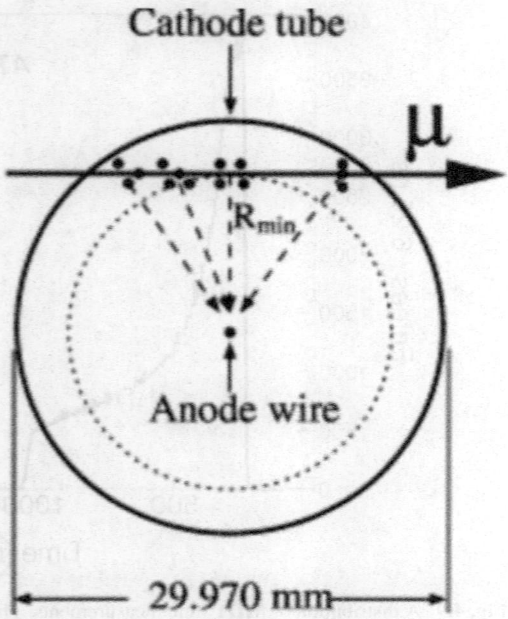

a transverse (ϕ) measurement with 5 mm resolution. The signals from the wires are not read out, rather a position is obtained by measuring the relative charge deposited on neighboring strips. The **CSC**s are segmented into chambers in ϕ, each consisting of four layers, resulting in four measurements per muon.

4.4.1.1 MDT Timing

The readout of the **MDT**s also gives a profile of the hits that compose the signal in time, seen in Fig. 4.9. The peak of the distribution, t_0, measures the time the signal began with respect to $t = 0$, the time of the pp interaction measured by the **ATLAS** clock. The end of the signal, $t_{\max}$, is the maximum drift time. In a perfectly calibrated detector, the t_0 measurement is Gaussian about $t_0 = 0$ with a width of about 3 ns.

4.4.2 Resistive Plate Chambers and Thin Gap Chambers

The trigger chambers must provide transverse momentum resolution in the barrel and endcaps over the full ϕ range; have time resolution such that detector information can be matched to the correct bunch crossing; be robust against photon and

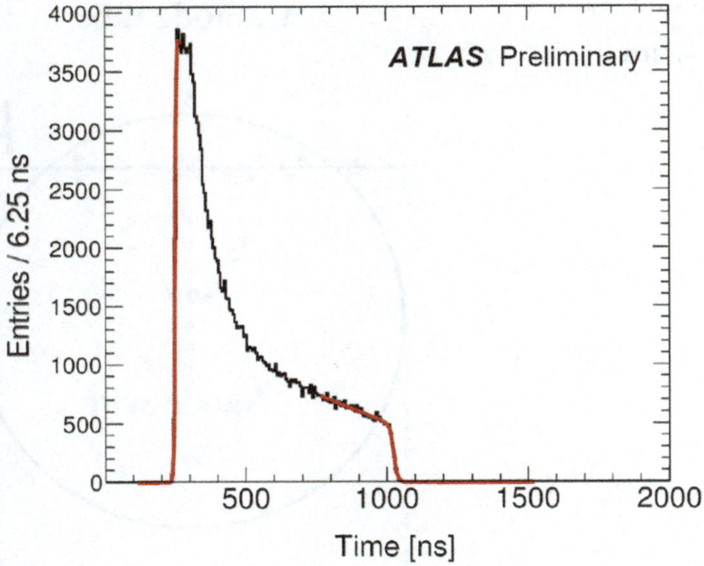

Fig. 4.9 A distribution of **MDT** time measurements. The t_0 is the time between 0 and the peak of the distribution, shown by the leftmost red line. The rightmost red line indicates t_{max}, the end of the **MDT** signal [10]

neutron backgrounds in the experimental cavern; and provide a ϕ measurement to complement the **MDT** η measurement.

In the barrel, **RPC**s are used to achieve these goals. An **RPC** is a parallel plate detector composed of two phenolic-melaminic plastic laminate plates 2 mm apart filled with $C_2H_2F_4$ gas and with a 4.9 kV/mm field between them. When a muon passes through the detector, it ionizes the atoms in the gas, the ionized electrons further ionize other atoms, creating an avalanche. Both sides of the chamber are readout and they have perpendicular readout strips, giving both z and ϕ measurements. The **RPC** are arranged into chambers of two detector layers each (giving two measurements per chamber) and positioned above or below each **MDT** chamber.

In the endcaps, multiwire proportional chamber **TGC**s are used. They work very similarly to the **CSC**s, with perpendicular anode wires and cathode strips, but the unit is more compact, giving the requisite increase in speed. Multiwire proportional chambers are ideal for the endcaps of the **MS** because they have an extremely granular readout, which allows them to deal with the inhomogeneity of the magnetic field in the transition region.

4.4.3 *Toroid Magnets*

The toroid magnet system, composed of a barrel and two end-caps, provides a toroidal magnetic field of 0.5 T and 1 T for the barrel ($|\eta| < 1.4$) and end-cap ($1.6 < |\eta| < 2.7$) regions of the **MS**. In the transition region ($1.4 < |\eta| < 1.6$) muons are bent by a combination of the two fields resulting in an extremely heterogeneous field making accurate measurement challenging. This inhomogeneity results in geometric regions where the muon trajectories are not bent at all, mimicking a high p_T muon, adding challenges for triggering on high p_T muons in this region. The toroid is much larger than the solenoid, 25.3 m long and 10 m in radial width, but also operates at a temperature of 4.5 K. All three toroid magnets are made of Al-stabilized Nb/Ti/Cu conductor. They have an air-core structure, which gives them a strong bending power over a large volume while minimizing material that could induce any additional scattering.

4.5 Particles in ATLAS

All of the **ATLAS** subdetectors are used in combination to identify particles. Charged particles interact with the **ID** resulting in hits to be reconstructed into tracks. A track that points to a cluster in one of the calorimeters indicates the kind of charged particle that made the track, and a cluster without an associated track indicates a neutral particle. The calorimeters are designed such that they absorb all of the energy of a particle and particles measured by the **EM** calorimeter should not enter the hadronic calorimeter, and hadronic particles should not enter the **MS**. Muons have minimal interactions with the calorimeters, but do leave a track in both the **ID** and **MS**. The only **SM** particle that escapes the detector entirely is a neutrino. An undetected particle, like an **SM** ν or some **BSM** particle, could be seen as an imbalance in transverse momentum referred to as Missing Transverse Energy (**MET**). The transverse momenta of all particles should sum to zero in order to conserve momentum, so any non-zero sum indicates an undetected particle (Fig. 4.10).

4.6 Trigger and Data Acquisition

The **LHC** provides bunch crossings every 25 ns in which an average of 33 collisions occur. Each bunch crossing, or *event*, contains about 25 MB of detector-level data. This can be reduced to about 1 MB of reconstructed data, but this would still result in 40 TB/s, an impossible amount of data to process and store. Furthermore, many of these events are not useful for physics analyses; most events contain low energy **QCD** process from marginal interactions between partons. A great deal of effort

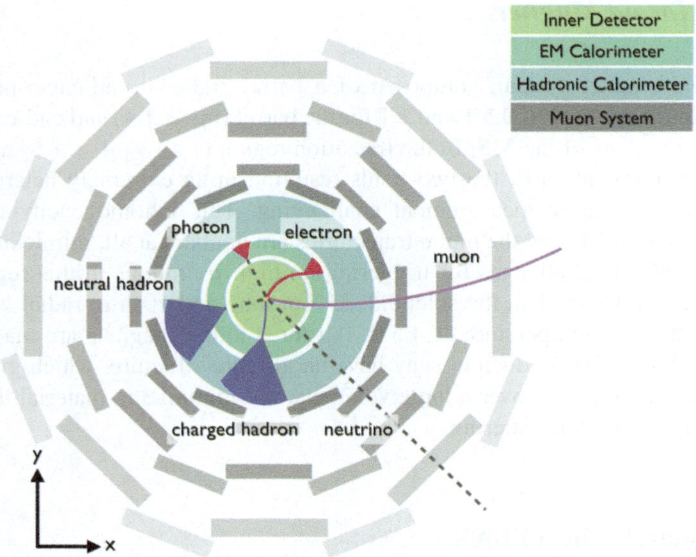

Fig. 4.10 A schematic of the signatures of Standard Model particles in the **ATLAS** detector that illustrates how the subdetectors are used together to identify particles. Dashed lines indicate a particle trajectory that leaves no detector signature. Figure not drawn to scale nor does it represent a real physical process

goes into the optimization of the algorithms used to select the events with the highest probability to be interesting for physics because if an event with some new **BSM** particle occurs in a collision, but is not recorded, it is lost forever.

4.6.1 Trigger Architecture

ATLAS uses multi-level trigger system to record 1000 events per second from the 40,000,000 per second delivered by the **LHC**. First, the hardware-based Level One (**L1**) trigger uses isolated detector information to make a decision in 2.5 μs and reduce the event rate from 40 MHz to 100 kHz. Then, the CPU-based High Level Trigger (**HLT**) uses higher level detector information from various subdetectors to make a final decision in 200 ms and further reduce the rate to 1 kHz. A schematic of the main components of the trigger system can be seen in Fig. 4.11.

The **L1** trigger has access to information from the calorimeters and the **MS** [11]. The **ID** is not used at this stage due to the large number of readout channels as well as the computational intensity of track reconstruction. L1 Calorimeter Trigger (**L1Calo**) uses information from both calorimeters to identify electrons, photons, τ leptons, jets, and **MET**. It divides the calorimeters into coarse-grained,

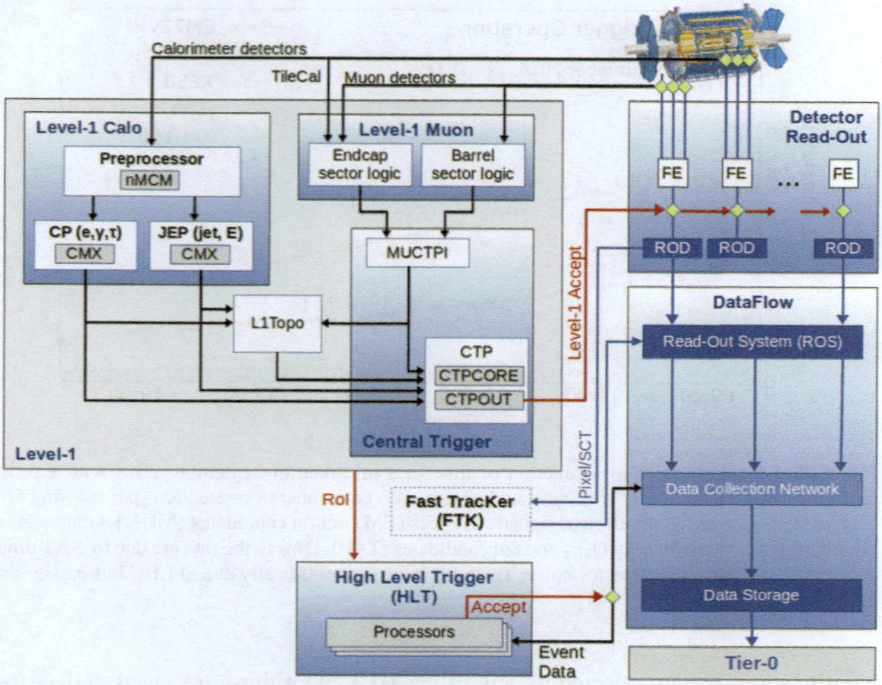

Fig. 4.11 A detailed schematic of the trigger and data acquisition infrastructure for Run 2. FTK was not a part of the Run 2 trigger system [11]

$\Delta\eta \times \Delta\phi = 0.1 \times 0.1$ towers. Its algorithms look for towers with large deposits to define an Region of Interest (**RoI**). Based on the shape and size of the summed energy deposits around the tower, the cluster is classified. It then correlates these clusters across the event, to identify the number of candidates of a given physics object, and calculates **MET**. L1 Muon Trigger (**L1Muon**) looks for coincidences between different layers of the **RPC** and **TGC** detectors to identify hight p_T muons consistent with the interaction point.

The decision to accept or reject an event at **L1** is made by the Central Trigger Processor (**CTP**), which uses a *trigger menu* that describes which event topologies with which energies should be passed to the **HLT**. If the event is accepted by the **L1** trigger, full information from the whole detector is read out and passed to the **HLT**. The **HLT** also receives the location of the **RoIs**. An example of the **L1** trigger rates in one run in September 2018 is shown in Fig. 4.12. About a third of events are selected as each an **EM**, muon, or hadronic signature.

The **HLT** receives much less data and therefore has more time to make a decision. The **HLT** has its own trigger menu with more complex topologies. Full event reconstruction cannot be performed in 200 ms, so each **HLT** algorithm is seeded by an **L1** trigger and tracking is performed in **RoIs** to refine the particle

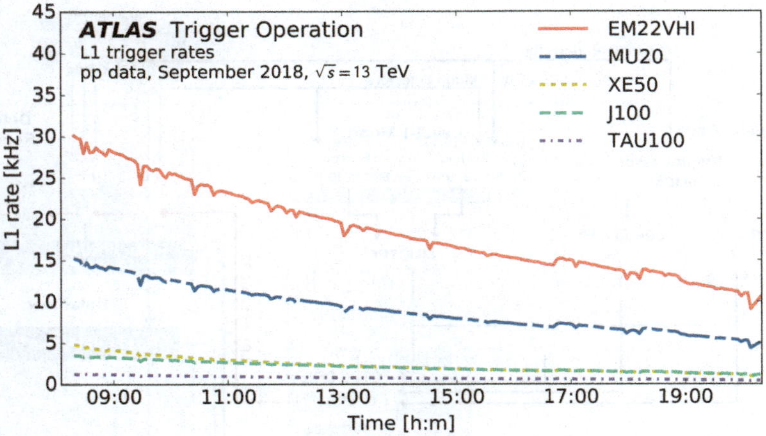

Fig. 4.12 **L1** trigger rates as a function of time in a fill taken in September 2018 with a peak luminosity of $L = 2.0 \times 10^{34}$ cm^{-2}s^{-1} and a peak average number of interactions per crossing 56. Trigger items are based on electromagnetic clusters (EM), muon candidates (MU), jet candidates (J), missing transverse energy (XE) and tau candidates (TAU). Dips in the rate are due to dead-time and spikes are caused by detector noise. The rates increase periodically due to LHC luminosity re-optimizations [12]

identification. Events selected by any of the **HLT** algorithms are saved to disk for offline reconstruction, described in the next chapter. An example of the **HLT** rates grouped by physical process in one run in September 2018 is shown in Fig. 4.13. The trigger rates represent collaboration priorities. Most events include jets, but the jet triggers do not take the largest fraction of the trigger rate. In fact, jets must be very high p_T or in events with other physics objects in order to be selected. Large proportions of the **HLT** rate is allocated to electrons and muons so that electroweak physics can be studied via events with leptonic W and Z boson decays. In general, single object triggers have very high p_T thresholds because they are more common, whereas multi-object triggers can select lower energy events while keeping the rate low.

Triggers are generally described by the physics object type and p_T threshold. Trigger rates can be limited by increasing the p_T threshold, or by making other selections on the object like limiting the η range or requiring isolation. Triggers have some *turn on* before they reach their peak efficiency, an example trigger is shown in Fig. 4.14. **HLT** triggers have higher energy thresholds than the **L1** triggers to ensure the algorithms are being run where the previous trigger is fully efficient. Offline selections are similarly made above the energy thresholds of the **HLT**.

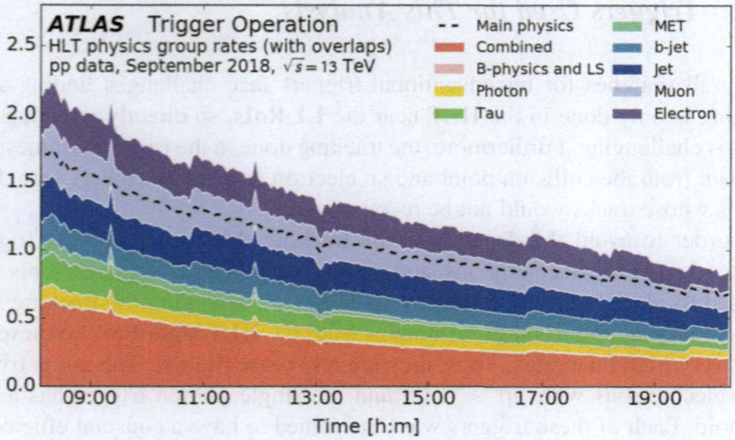

Fig. 4.13 HLT trigger rates grouped by physics process as a function of time in a fill taken in September 2018 with a peak luminosity of $L = 2.0 \times 10^{34}$ cm^{-2}s^{-1} and a peak average number of interactions per crossing 56. Each physics group contains single-object and multi-object triggers. The "combined" combined represents triggers with multiple kinds of physics objects, like combinations of electrons, muons, taus, jets and **MET**. All rates decrease exponentially with decreasing luminosity over the course of the fill. Dips in the rate are due to dead-time and spikes are caused by detector noise. The rates increase periodically due to LHC luminosity re-optimizations [12]

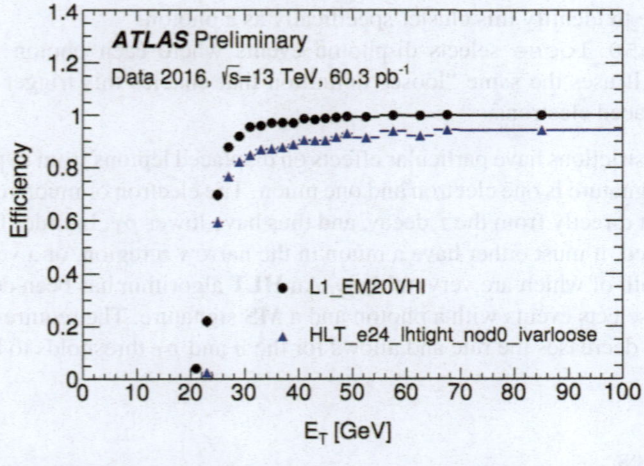

Fig. 4.14 An example of trigger efficiency as a function of transverse energy, E_T, for a electron trigger. The **L1** trigger is shown in black and the **HLT** algorithm it seeded is shown in blue [13]

4.6.2 Triggers Used for This Analysis

Nearly all searches for unconventional triggers face challenges finding a trigger. Tracking is only done in the **HLT** near the **L1 RoIs**, so directly targeting peculiar tracks is challenging. Furthermore, the tracking done in the trigger assumes particles originate from the collision point and so electron and muon triggers veto displaced leptons whose tracks would not be reconstructed.

In order to avoid the dependence on tracking, **MS**-signature-only triggers are used to identify displaced muons, and photon triggers (**EM**-signature-only triggers) are used to identify displaced electrons. However, removing the tracking information removes an important discriminator from the **HLT** algorithm, so these triggers suffer very high fake rates. Thus, they are otherwise limited. The muon trigger can only select muons with $|\eta| < 1.05$ and the single photon trigger has a high p_T threshold. Each of these triggers was determined to have a constant efficiency with respect to the displacement of the lepton. The triggers used for this analysis are listed below and the requirements made on them in the analysis will be described in Sect. 6.1.

- `HLT_mu60_0eta105_msonly` selects single **MS**-only muon signatures with $p_T > 60\,\text{GeV}$. The η range is restricted to $|\eta| < 1.05$ to reduce the fake rate.
- `HLT_g140_loose` selects single photons, in this analysis used as **EM**-only signatures. The fake rate for this trigger is quite high, so the photons must have $E_T > 140\,\text{GeV}$. "loose" refers to the fact that very few additional requirements are made to identify this cluster specifically as a photon.
- `HLT_2g50_loose` selects di-photon events where each photon has $E_T > 50\,\text{GeV}$. It uses the same "loose" definition that enables this trigger to be used for displaced electrons.

These restrictions have particular effects on displaced leptons from $\tilde{\tau}$, particularly when the signature is one electron and one muon. The electron or muon result from a τ decay, not directly from the $\tilde{\tau}$ decay, and thus have lower p_T. In order for an event to be selected, it must either have a muon in the narrow η region, or a very high p_T electron, both of which are very unlikely. An **HLT** algorithm has been designed for Run 3 that selects events with a photon and a **MS** signature. The requirement of the two objects decreases the rate and allows for the η and p_T thresholds to be relaxed.

References

1. ATLAS Collaboration, The ATLAS experiment at the CERN Large Hadron Collider. JINST **3**, S08003 (2008). https://doi.org/10.1088/17480221/3/08/S08003
2. CERN, Detector and Technology (2020). https://atlas.cern/discover/detector
3. D. Bortoletto, How and why silicon sensors are becoming more and more intelligent? J. Instrument. **10**(08), C08016–C08016 (2015). ISSN: 1748-0221. https://doi.org/10.1088/1748-0221/10/08/c08016

4. ATLAS Experiment, ATLAS Inner Detctor (2014)
5. ATLAS Experiment, Inner Detector (2014). https://cds.cern.ch/images/CERN-GE-0803014-01/file?size=medium
6. ATLAS Collaboration, Topological cell clustering in the ATLAS calorimeters and its performance in LHC Run 1. Eur. Phys. J. C **77**, 490 (2017). https://doi.org/10.1140/epjc/s10052-017-5004-5. arXiv: 1603.02934 [hep-ex]
7. N. Nikiforou, Performance of the ATLAS Liquid Argon Calorimeter after three years of LHC operation and plans for a future upgrade. Technical report. arXiv:1306.6756. Comments: 12 pages, 25 figures, Proceedings of talk presented in 2013. https://cds.cern.ch/record/1558820
8. T. Hebbeker, K. Hoepfner, Muon spectrometers, in *Handbook of Particle Detection and Imaging*, ed. by C. Grupen, I. Buvat (Springer, Berlin, 2012), pp. 473–496. ISBN: 978-3-642-13271-1. https://doi.org/10.1007/978-3-642-13271-1_19
9. ATLAS Collaboration, Standalone vertex finding in the ATLAS muon spectrometer. JINST **9**, P02001 (2014). https://doi.org/10.1088/1748-0221/9/02/P02001. arXiv: 1311. 7070 [hep-ex]
10. ATLAS Experiment, Public Muon Spectrometer Plots for Collision Data (2012)
11. ATLAS Collaboration, Performance of the ATLAS trigger system in 2015. Eur. Phys. J. C **77**, 317 (2017). https://doi.org/10.1140/epjc/s10052-017-4852-3. arXiv: 1611.09661 [hep-ex]
12. ATLAS Experiment, Trigger Operation Public Results (2019). https://twiki.cern.ch/twiki/bin/view/AtlasPublic/TriggerOperationPublicResults
13. ATLAS Experiment, Public Egamma Trigger Plots for Collision Data

Chapter 5
Object Reconstruction

Event reconstruction is the process by which detector signals are converted into objects that can be used for physics analysis. This is a complex process that requires a great deal of focused effort by the **ATLAS** collaboration. First, digital signals from the detector are collected into tracks and clusters, then they are combined to form reconstruction-level physics objects. Then, a identification steps is performed, where quality requirements are placed on the reconstruction-level objects to identify them as signatures of physical particles, like electrons and muons, that can be used in physics analyses.

These algorithms are centrally developed by the collaboration and are generally designed to reconstruct and identify prompt objects. This section describes this process for objects which are relevant to this analysis, as well as the changes to these algorithms that have been implemented to be able to study displaced leptons.

This chapter describes the reconstruction of objects on which selections are made in the analysis: electrons, muons, and their tracks. This is only a subset of the physics objects used by other **ATLAS** searches, which are not described here. Jets and photons are reconstructed, but no selections are made on them.

5.1 Track Reconstruction

Track reconstruction is the process by which **ID** data is converted into particle trajectories. This is a complicated process, due to both the density of each event in the detector as well the helical trajectory (in ϕ) the particles take due to the solenoidal magnetic field. Tracks are described by five parameters with respect to the beamspot position: d_0 (the transverse point of closest approach, or transverse impact parameter), z_0 (the longitudinal point of closest approach, or longitudinal impact parameter), ϕ (the azimuthal angle of the track momentum), θ (the polar angle of the track momentum), and q/p (the ratio of the track's charge to the magnitude of its

L. Horyn, *A Search for Displaced Leptons in the ATLAS Detector*, Springer Theses,
https://doi.org/10.1007/978-3-030-91672-5_5

momentum) [1]. This section describes the standard track reconstruction algorithm used by all **ATLAS** searches, as well as the extended algorithm, Large Radius Tracking (**LRT**), used to identify displaced objects.

5.1.1 Tracking Algorithms

5.1.1.1 Kalman Filter

Kalman filters are widely used across LHC experiments for track fitting. It is a recursive algorithm that allows the user to efficiently extrapolate from a track seed. In the sequential, or extended, Kalman filter, uses a linear approximation to the track path. It makes a prediction for the next layer of detector material based on the seed, and then looks for a hit in that region. At each step, it updates the linearization to improve the measurement for the next step, where each measurement is weighted by its certainty. **LRT** uses this approach.

In Standard Tracking (**ST**), a combinatorial Kalman filter is used. This method employs several Kalman filters running in parallel and allows for the assumption that the measurement in the next layer is not necessarily assumed to be part of the track that formed the seed (as is often the case in the dense environment of the **ATLAS ID**). At each successive layer in track finding, several branches are extended if several measurements can be found in the same layer. The update of the linearization is done independently for each branch, and branches are created for missing measurements to account for detector inefficiencies. The branch is extended to the next layer if a measurement is found, and the process is repeated. Branches with no measurement for several layers are removed, and at the end, the branch with the best quality is selected. This process allows for parallelization of a complex combinatorial process [2].

The Kalman filter works well for tracking in environments such as **ATLAS** because even though it is computationally intensive, it generally gives the best precision.

5.1.1.2 Hough Transform

A Hough transform is a pattern recognition algorithm that is useful for identifying a particular signature that can be described in a known parametric form. It performs well in noisy environments and is tolerant of holes in the signature. In relies on a transformation from physical to parameter space. Each point in physical space maps to a line in parameter space, and the intersection of the parameter space lines gives the values of constants of the parametric form.

For example, if the pattern being searched for is a line described by $y = mx + b$, a Hough transform maps from $x - y$ space to $m - b$ space. Each point (x_i, y_i) in the original image maps to a line in parameter space. If all (x_i, y_i) are plotted together,

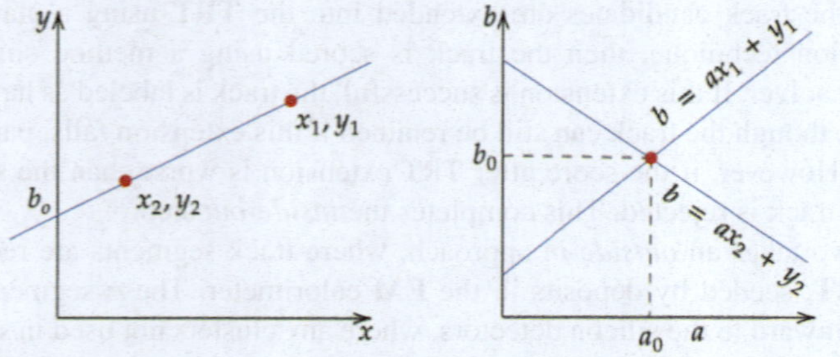

Fig. 5.1 An illustration of $x - y$ space (left) and $m - b$ space used to find lines via the Hough transform [4]

the intersections of the lines in parameter space at (m_i, b_i) define lines in physical space with parameters m_i and b_i [3]. This is sketched in Fig. 5.1.

The Hough transform is a *pattern finding* algorithm, not a tracking algorithm. It is not ideal for full **ID** tracking, because its complexity grows exponentially with the number of dimensions and the form of the pattern must be known *a priori*. It is, however, very useful for reconstructing **MS** segments.

5.2 Standard Tracking

First, clusters from the **ID** are converted into three-dimensional *space-points*. In the Pixel detector, a space-point is simply one hit, while in the **SCT**, hits from both layers of the sterostrip form a three-dimensional measurement.

Tracking in ATLAS is performed in two steps. During the first step, called *inside-out*, tracks are seeded from the silicon layers. In the Pixel and **SCT** detectors, track seeds are formed from sets of three space-points, each from a separate silicon layer. If the seed passes an assortment of selection criteria, including on the p_T and d_0, track candidates are built using a combinatorial Kalman filter (further described in Sect. 5.1.1.1). Seed requirements serve to cut down on the number of times the computationally expensive Kalman filter must be run. Multiple track candidates can be built from the same seed.

Since all possible track combinations are created in the previous stage, an *ambiguity solving* step is now required. Tracks are scored based on a variety of criteria, the number of holes (active sensor material with which the track would have intersected and do not contain a cluster), χ^2 (to prioritize tracks with a better fit), and p_T (to prioritize tracks with a higher p_T). A further requirement that no more than two tracks may share the same cluster reduces the number of duplicate tracks; however, a cluster may be removed from a track to stay within this limit and the track is then re-scored.

Next, the track candidates are extended into the **TRT** using a classical track extrapolation technique, then the track is scored using a method similar to the ambiguity solver. If this extension is successful, the track is labeled as having a **TRT** extension, though the track can still be retained if this extension fails, particularly at large $|\eta|$. However, if the score after TRT extension is worse than the silicon-only score, the track is rejected. This completes the *inside-out* step.

Step two takes an *outside-in* approach, where track segments are reconstructed in the **TRT**, seeded by deposits in the **EM** calorimeter. These segments are then extended inward to the silicon detectors, where any clusters not used in step one can be associated to the track. The extension inward is not required, as **TRT**-only tracks are used for reconstructing and identifying converted photons.

5.2.1 Large Radius Tracking

LRT [5] is required to reconstruct tracks with impact parameters larger than what is allowed by the **ST** algorithm. These requirements are designed for a maximum displacement a few mm, to enable the identification of heavy flavor hadron decays. Standard tracking cuts off at $|d_0| = 10\,\text{mm}$, and the range $10\,\text{mm} < |d_0| < 300\,\text{mm}$ is covered by **LRT**.

An optional third step of the tracking chain, it uses the same *inside-out* tracking algorithm, but relaxes various requirements that allow for a much more inclusive track collection. The major changes are summarized in Table 5.1. The performance of the different algorithms are compared in Figs. 5.2 and 5.3. These cuts are applied during both the seeding and extension steps. Additionally, **LRT** uses a sequential Kalman filter as opposed to the combinatorial Kalman filter used in **ST**.

LRT is required for this analysis, but cannot be applied to all events in the Run 2 dataset. The full event reconstruction with **LRT** takes about 2.5 times longer than with **ST**, so events are filtered based on the triggers that selected them, such that this algorithm is only run on about 10% of the dataset. This filtering is further described in Sect. 6.1.

Table 5.1 Most important cuts that differ between **ST** and **LRT**

Cut	**ST**	**LRT**
d_0^{max} (mm)	10	300
z_0^{max} (mm)	250	1500
$\lvert\eta^{\text{max}}\rvert$	2.7	5
Max shared silicon modules	1	2
Min unshared silicon clusters	6	5
Min number of silicon hits	7	7

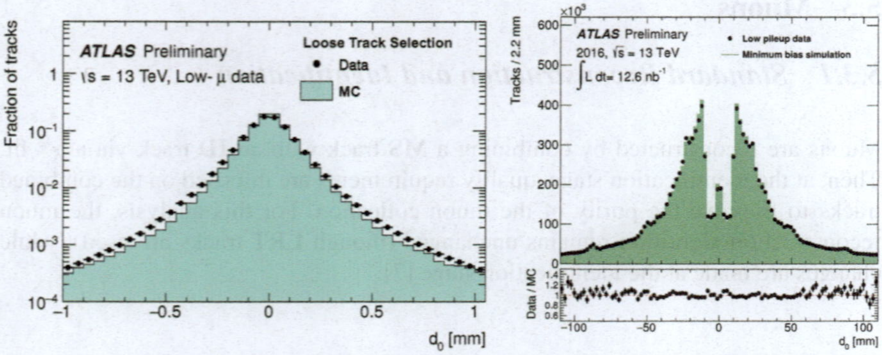

Fig. 5.2 Number of tracks reconstructed with respect to d_0 in **ST** (left) [6] and **LRT** [5]. Note the difference in x-axis range

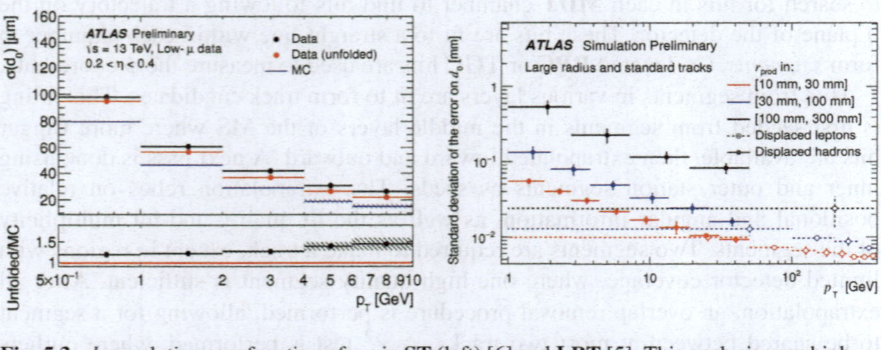

Fig. 5.3 d_0 resolution as a function of p_T in **ST** (left) [6] and **LRT** [5]. This analysis uses high p_T leptons with high p_T tracks, and thus with very good d_0 resolution

5.2.2 *Primary Vertex Reconstruction*

After all of the tracks are reconstructed, they must be correctly assigned to a Primary Vertex (**PV**). A **PV** is the point in space where the *pp* collision occurred. Generally, there are many **PV**s per event: one is the hard-scatter, high-energy event of interest, and the others are pileup. In the Run 2 dataset, there are an average of 33 **PV**s per event.

PVs are reconstructed using an iterative vertexing procedure. First, good quality tracks, combined with the beamspot measurement, are used to find an optimal vertex position. Then each track is assigned a weight based on its compatibility with that position, and the vertex position is then recomputed using the weights of the tracks. After the final vertex position is determined, tracks very incompatible with the vertex are removed from it and can be used to create another vertex. Any vertex with at least two tracks are considered **PV**s.

5.3 Muons

5.3.1 Standard Reconstruction and Identification

Muons are reconstructed by combining a **MS** track with an **ID** track via a χ^2 fit. Then, at the identification stage, quality requirements are imposed on the combined tracks to improve the purity of the muon collection. For this analysis, the muon reconstruction algorithm remains unchanged (though **LRT** tracks are used), while changes are made at the identification stage [7].

5.3.1.1 MS Track Reconstruction

To reconstruct **MS** tracks, a Hough transform (described in Sect. 5.1.1.2) is used to search for hits in each **MDT** chamber to find hits following a trajectory on the η plane of the detector. These hits are fit to a straight line within each chamber to form *segments*. Co-located **RPC** or **TGC** hits are used to measure the ϕ coordinate.

Hits from segments in various layers are fit to form track candidates. This fitting is first seeded from segments in the middle layers of the **MS** where more trigger hits are available, then extrapolated inward and outward. A next pass is done using inner and outer station segments as seeds. The extrapolation relies on relative positional and angular information, as well as the fit quality and hit multiplicity of the segments. Two segments are required to make a track, except in regions with limited detector coverage, where one high quality segment is sufficient. After all extrapolation, an overlap removal procedure is performed, allowing for a segment to be shared between at most two tracks. A χ^2 test is performed, where outliers hits can be removed from the track and additional hits consistent with the track candidate's trajectory can be added.

5.3.1.2 Combined Muons

Next, the **MS** tracks are combined with **ID** tracks to form Combined (**CB**) muons, a track fit over the two tracks. **CB** muon tracks are generally seeded from the **MS**, then extrapolated inward and matched to an **ID** track, but the inverse is also allowed. Hits from the **MS**, not the **ID**, may be added or removed to improve the fit between the two tracks.

5.3.1.3 Muon Identification

This analysis uses the default muon working point for ATLAS analyses, called a *medium* muon. This working point places a requirement on the number of **ID** and **MS** hits that comprise the track to ensure a robust momentum measurement. At

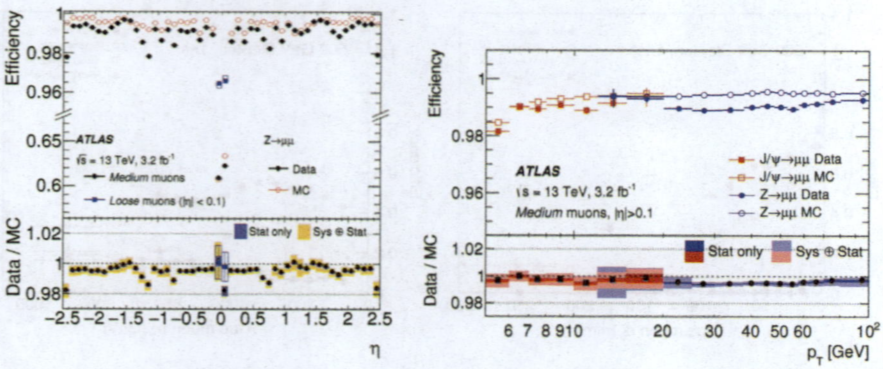

Fig. 5.4 Medium muon identification efficiency with standard tracking and criteria. Left is efficiency vs η and right is efficiency as a function of p_T. Medium muons are reconstructed very efficiently, except for around $|\eta| \approx 0$, where the **MS** is missing detector coverage [7]

least 1 pixel hit, at least 5 **SCT** hits are required and at least 10% of the **TRT** hits associated to the object must be included in the final fit. There must also be fewer than 3 holes in the silicon tracking layers. Furthermore, the **MS** track must have at least 3 hits in at least 2 s**MDT** layers. In the crack region $|\eta| < 0.1$, **MS** tracks with at least three hits in only one **MDT** layer are allowed provided there are no holes in the track. Finally, a loose requirement is placed on the consistency between the **MS** and **ID** tracks. Namely, the *q/p significance*, the difference between the charge and momentum ratio in the **ID** and **MS** divided by their uncertainties summed in quadrature, is required to be less than 7. The performance of the muon reconstruction and identification algorithms with standard tracking are shown in Fig. 5.4.

5.3.2 *Modifications*

For this analysis, muons are reconstructed after **LRT** is performed and the reconstruction and identification efficiency is quite high. Furthermore, at the identification stage, the requirement that the **ID** track has at least one pixel hit is removed, further improving the efficiency at high d_0. The effect of these improvements is show in Fig. 5.5.

5.4 Electrons

Electrons are reconstructed using clusters from the **EM** calorimeter as well as tracks from the **ID** [8]. Electron reconstruction brings more complication and ambiguity than muon reconstruction because of the presence of photons, converted photons,

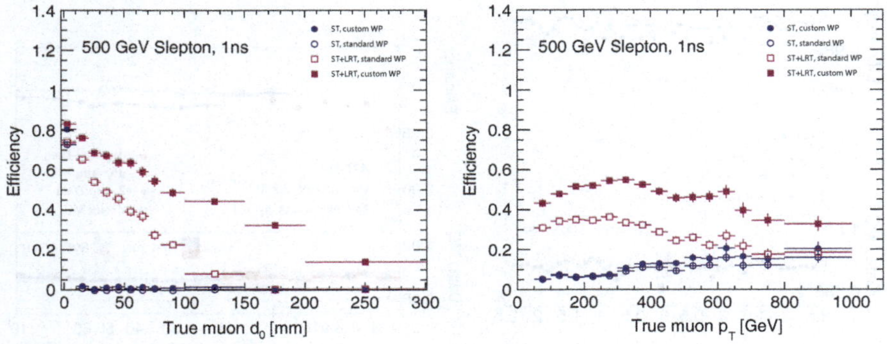

Fig. 5.5 Muon identification efficiency with modified criteria. Left is efficiency vs d_0 and right is efficiency as a function of p_T. The blue denotes reconstruction with **ST**, and the pink with **LRT**, and the filled in markers use the modified identification working point

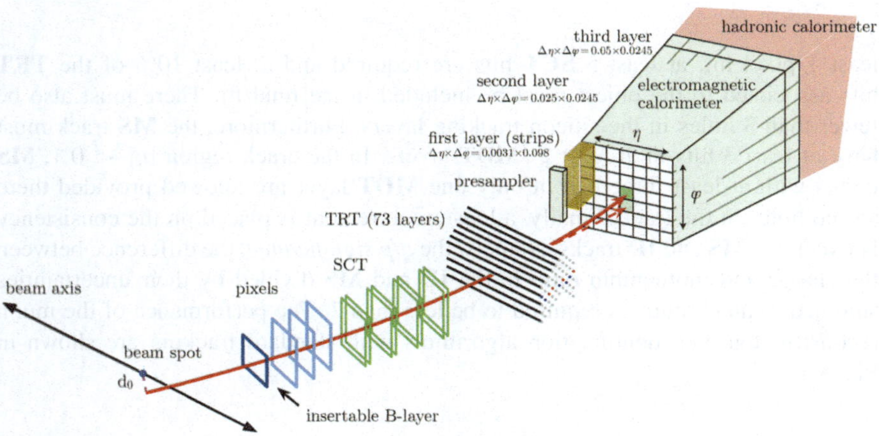

Fig. 5.6 A sketch of an electron's path through a slice of the **ATLAS** detector [8]

and bremsstrahlung radiation from electrons moving through material. An electron's path through the ATLAS detector can be seen in Fig. 5.6. These factors make the identification and accurate measurement of electrons quite challenging. More than in muon reconstruction, it relies on displacement-based quality information, posing a problem for this search. In order to select electrons with high d_0, some of these selections are modified and electrons are reconstructed with the **LRT** collection. However, due to the more challenging reconstruction, the electron selection efficiency is lower than that for muons.

5.4.1 *Standard Reconstruction and Identification*

5.4.1.1 Cluster Reconstruction

First, clusters are formed from $\eta \times \phi$ *towers* of size $\Delta\eta \times \Delta\phi = 0.025 \times 0.025$ (roughly granularity of the second layer of the **EM** calorimeter, where about 80% of the energy in a shower is deposited). In each region, the energy deposited in all layers of the calorimeter is summed and are used as input to a seeding algorithm to form clusters.

A *sliding window* algorithm [9] is used to form clusters. In this algorithm, an 3×5 window is slid across each tower, the energy is summed inside of this window. If the sum is a local maximum and is above a threshold of $E_{\mathrm{T}} > 2.5$ GeV, this window is considered a *cluster*. A duplicate removal process is then performed for nearby clusters with similar energy measurements, keeping only the cluster with the largest E_T. Inefficiency in the cluster reconstruction step is negligible compared to the uncertainty in the next two steps. The efficiency of this step is 65% at $E_{\mathrm{T}} = 4.5$ GeV and >99% above $E_{\mathrm{T}} = 15$ GeV.

5.4.1.2 Tracking

Since electrons are so light, they can lose a significant amount of their energy due to bremsstrahlung radiation as they traverse the **ID**, thus resulting in a track seed that cannot be extended to the requisite number of silicon layers using the processes described in Sect. 5.1. Thus, a second pass at tracking, allowing for 30% energy loss due to bremsstrahlung radiation at each detector layer is performed in the vicinity of a good quality **EM** cluster. Here, the track candidate p_T is lowered to 400 MeV (compare to 1 GeV), but still uses the standard hypothesis that the track is like that from a pion and does not emit substantial bremsstrahlung radiation. If this fit still fails, a third pass is performed using the assumption that the track is like that from an electron, allowing for an additional degree of freedom in the χ^2 calculation that accounts for additional radiation. In all, this gives 98% reconstruction efficiency for electrons with $E_{\mathrm{T}} > 10$ GeV.

This loosened track fitting requirements allow for increased efficiency, but the resulting tracks do not correctly account for the energy loss of the electron to the material. An additional tracking pass, using an optimized Gaussian-sum Filter (**GSF**) is used to correct for this.

This procedure is performed on tracks with at least 5 silicon hits and roughly match to an **EM** cluster. The **GSF** procedure is based on a combinatorial Kalman filter, resulting in track parameters weighted by Gaussian function describing material-induced energy losses. This procedure also accounts for the increase in curvature caused by the decrease in momentum due to energy loss in material, improving the calculation of track parameters. An example of this is shown in Fig. 5.7. The reconstruction efficiency for this step is around 98% for electrons with $E_{\mathrm{T}} > 30$ GeV.

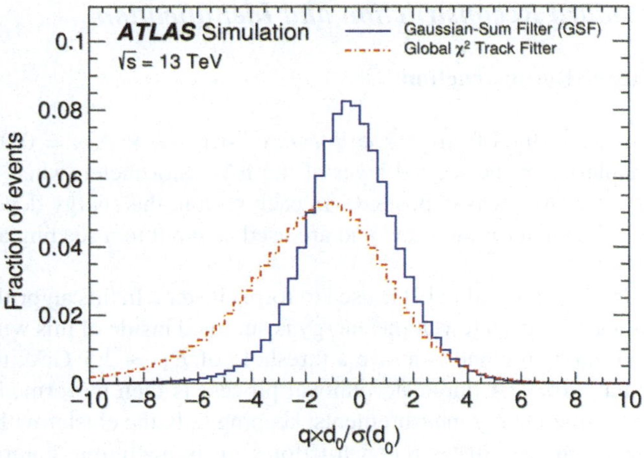

Fig. 5.7 The impact on the $q \times d_0/\sigma(d_0)$ measurement by the addition of **GSF** tracking. This is sampled from prompt electrons, so the narrower distribution centered around zero indicates significant improvement [8]

5.4.1.3 Combined Reconstruction

To put it all together, **GSF** tracks are then matched to seed calorimeter clusters then the final cluster size is determined. Tracks and clusters are matched with stricter requirements, requiring that their ϕ measurement is within $+0.05$ or -0.10. It is possible that several tracks might match to the same cluster, but a primary track is assigned based on its proximity and quality. If the track can be associated to a secondary vertex, it is classified as a potential photon conversion, not an electron. Electrons are further distinguished from photon conversions using their E/p, p_T, and number of pixel hits.

Final clusters are formed by looking in a window around the seed cluster. The reconstructed electron's energy measurement is taken from its full cluster, and directional information taken from its track. At $E_{\mathrm{T}} > 15$ GeV, there is a 99% efficiency to reconstruct an electron (provided it has at least one pixel hit and at least seven total silicon hits on track). The performance of the electron reconstruction and identification algorithms with standard tracking are shown in Fig. 5.8.

5.4.1.4 Identification

Electrons are identified using a likelihood that further distinguishes them from photons, light flavor jets, and leptonic heavy flavor decays. The likelihood is more flexible than a simple cut-based method, allowing for electrons to fail one selection criterion. It also allows for the use of discriminating criteria which have relatively similar shapes in signal and background. A some cuts are made at the identification

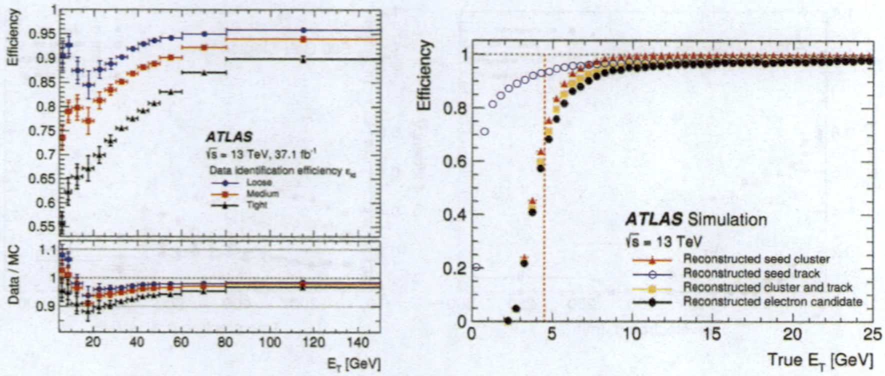

Fig. 5.8 Electron reconstruction efficiency as a function of true transverse energy, E_T. For the various electron working points (left) and for each step in the reconstruction process (right) [8]

step, including on the number of pixel and silicon hits, as well as on the shower width. Many other factors are not cut on but are included in the likelihood including the track quality, consistency between the electron's track and its cluster, energy ratios in the various layers of the **EM** calorimeter, and the **EM** calorimeter compared to the hadronic calorimeter.

5.4.2 Modifications

To be able to reconstruct electrons with high impact parameter, several changes needed to be made to the reconstruction and identification algorithms.

Similar to the muon case, the reconstruction is then run on the track collection including **LRT** tracks. At the identification stage, variables concerned with d_0 are removed from the likelihood consideration, but do not retrain the likelihood itself. The standard algorithm requires 7 silicon hits and 1 pixel hit, and these requirements are removed in the modified algorithm. The effect of these improvements is shown in Fig. 5.9.

5.5 Overlap Removal

Each of the reconstruction chains described above, as well as those to reconstruction τ leptons, jets, and other physics objects used in **ATLAS**, are run simultaneously. This means that the same track or cluster can be used to reconstruct two different types of physics objects. An iterative process of Overlap Removal (**OR**) is used to

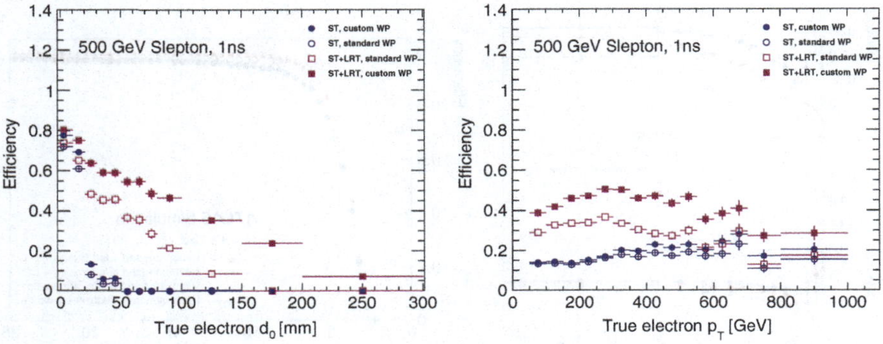

Fig. 5.9 Muon identification efficiency with modified criteria. Left is efficiency vs d_0 and right is efficiency as a function of p_T. The blue denotes reconstruction with **ST**, and the pink with **LRT**, and the filled in markers use the modified identification working point

remove such artifacts. Overlap removal is performed on baseline objects, further described in Chap. 9, which only pass identification and a loose p_{T} cut.

Electrons $\Delta R < 0.01$ from a muon are removed to suppress contributions from muon bremsstrahlung radiation. Then, any baseline jet residing within $\Delta R < 0.2$ of a baseline electron is removed. Baseline electrons with $E_T < 50$ GeV ($E_T > 50$ GeV) within $\Delta R < 0.4$ ($\Delta R < \min(0.4, 0.04 + 10\ \text{GeV}/p_{\mathrm{T}})$) of any remaining jet are removed. The overlap removal procedure between muon and jet candidates is designed to remove muons that may originate from the decay of hadrons, with the overlapping jet being retained. Nearby jets may also occur as a result of high-p_{T} bremsstrahlung, in which case the jet should be removed and the muon retained. Such jets are removed if they reside within $\Delta R < 0.2$ of a baseline muon and have less than three matched inner detector tracks. Muons are removed if they reside within either $\Delta R < 0.4$ or $\Delta R < \min(0.4, 0.04 + 10\ \text{GeV}/p_{\mathrm{T}})$ of any remaining jet, following the same p_{T}-dependent criteria as outlined for electrons above.

5.6 Isolation

In this analysis, both muons and electrons are required to be isolated to reduce background from heavy flavor decays. That is, they are not expected to be surrounded by much other activity from the hard scatter in either the Inner Detector or Calorimeters. Isolation is generally measured in terms as the scalar sum of energy in a radius $\Delta R = \sqrt{(\Delta\eta)^2 + (\Delta\phi)^2}$ around the lepton. Track based isolations are called p_T^{varcone} and calorimeter based isolations called E_T^{topocone}.

For example, the track-based isolation called $p_T^{\text{varcone30}}$ is the scalar sum of the transverse momenta of tracks with $p_T > 1$ GeV in a cone of $\Delta R = \min(10\ \text{GeV}/p_T^{\ell}, 0.3)$, then a value is selected to determine what should be

Table 5.2 Isolation cuts for electrons and muons

Variable	Electron cut	Muon cut
`ptvarcone20/pt`	<0.06	<0.04
`topoetcone20/pt`	<0.06	<0.15

considered "isolated". Whereas the calorimeter-based isolation counts the energy in a specified ΔR. For both calorimeter and **ID** isolation, requirements are placed on the value of the isolation variable divided by the p_{T} of the object. Isolation working points are centrally defined as a combination of cuts on the track-based and calorimeter-based isolation.

These definitions are much simpler for muons than for electrons. Electrons are very likely to emit bremsstrahlung radiation as they traverse the inner detector and those photons can then convert back into electrons. The tracks from these secondary particles are considered part of the electron's p_{T}. Furthermore, the electron leaves its own energy deposit in the **EM** calorimeter and this energy must be subtracted from the E_T^{topocone} calculation.

In this analysis the `FCTight` isolation definition is used for both electrons and muons, detailed in Table 5.2. In general, the requirements are stricter for electrons due to the identification ambiguity.

References

1. ATLAS Collaboration, The Optimization of ATLAS Track Reconstruction in Dense Environments ATL-PHYS-PUB-2015-006 (2015). https://cds.cern.ch/record/2002609
2. R.G. Brown, P.Y.C. Hwang, *Introduction to Random Signals and Applied Kalman Filtering with Matlab Exercises*, 4th edn. (Wiley, Hoboken, 2012)
3. M. Mårtensson, Fast pattern recognition with the ATLAS L1Track trigger for the HL-LHC, in *The 25th International Workshop on Vertex Detectors* (La Biodola, Isola d'Elba, ITALY) (2016)
4. Towards data science, Lines Detection with Hough Transform (2020). https://towardsdatascience.com/lines-detection-with-hough-transform-84020b3b1549
5. ATLAS Collaboration, Performance of the reconstruction of large impact parameter tracks in the inner detector of ATLAS. ATL-PHYS-PUB-2017-014 (2017). https://cds.cern.ch/record/2275635
6. ATLAS Experiment, InDetTrackingPerformanceApprovedPlots (2012). https://twiki.cern.ch/twiki/bin/view/AtlasPublic/InDetTrackingPerformanceApprovedPlots
7. ATLAS Collaboration, Muon reconstruction performance of the ATLAS detector in proton–proton collision data at $\sqrt{s} = 13$ TeV. Eur. Phys. J. C **76**, 292 (2016). https://doi.org/10.1140/epjc/s10052-016-4120-y. arXiv: 1603.05598 [hep-ex]
8. ATLAS Collaboration, Electron reconstruction and identification in the ATLAS experiment using the 2015 and 2016 LHC proton–proton collision data at $\sqrt{s} = 13$ TeV. Eur. Phys. J. C **79**, 639 (2019). https://doi.org/10.1140/epjc/s10052-019-7140-6. arXiv: 1902.04655 [hep-ex]
9. W. Lampl et al., Calorimeter Clustering Algorithms: Description and Performance. ATL-LARG-PUB-2008-002 (2008). https://cds.cern.ch/record/1099735

Chapter 6
Data and Monte Carlo Samples

Data and simulated Monte Carlo samples are used together to perform this analysis. ATLAS data from Run 2 of the LHC undergoes a special processing in order to make use of the augmented tracking and object reconstruction required to identify displaced leptons. Monte Carlo samples of both the benchmark model and representative Standard Model backgrounds are generated to define backgrounds and make a interpretation of the results in a specific signal model. All backgrounds are estimated from data, but Monte Carlo is used to optimize signal selection and understand properties of backgrounds.

6.1 Data

This analysis makes use of 139 fb^{-1} of $\sqrt{s} = 13$ TeV *pp* collision data delivered by the **LHC** and taken by the **ATLAS** detector between 2015 and 2018. The data used in this analysis were collected using three different triggers depending of the topology of the event, described in Table 6.1. They are applied sequentially in the order.

The **LRT** and special reconstruction used in this analysis are computationally intensive and introduce many more fake tracks and thus physics objects, so it is only run over a subset (roughly 10%) of the full `physics_Main` dataset. Events are filtered based on which triggers they fired as well as loose requirements on reconstruction-level objects created with the standard tracking algorithm. This analysis uses filters based on the triggers described in Table 6.1 combined with selections on combined muons, **MS**-only muons, electrons, and photons (to account for loss of efficiency reconstructing electrons with standard tracking) listed in Table 6.2.

If an event is selected by a filter, it is saved in its `RAW`, or detector-level, format in a `DRAW_RPVLL` dataset. This `DRAW_RPVLL` dataset then undergoes

L. Horyn, *A Search for Displaced Leptons in the ATLAS Detector*, Springer Theses,
https://doi.org/10.1007/978-3-030-91672-5_6

Table 6.1 Summary of trigger selection based on event topology. The first line is requested, if the event topology is not met, the second line is requested. If the event has the specified topology, but fails the trigger selection, the event is not selected

Topology	Trigger
`if` ≥ 1 electron, $p_T > 160$ GeV	`HLT_g140_loose`
`else if` ≥ 2 electrons, $p_T > 60$ GeV	`HLT_2g50_loose`
`else if` ≥ 1 muon, $p_T > 60$ GeV, $\lvert\eta\rvert < 1.07$	`HLT_mu60_0eta105_msonly`

Table 6.2 Selection applied to select single e/γ or muon event objects for `DRAW_RPVLL` processing. For muons, the $\lvert d_0\rvert > 2$ mm requirement is only enforced on combined muons (not **MS**-only muons which do not measure $\lvert d_0\rvert$)

	Object 1				Object 2			
Filter		p_T [GeV]	$\lvert\eta\rvert$	$\lvert d_0\rvert$ [mm]		p_T [GeV]	$\lvert\eta\rvert$	$\lvert d_0\rvert$ [mm]
Single muon	μ	>60	<1.07	>2				
Single photon	γ	>160	<2.5	–	γ	>10	<2.5	–
	γ				e	>10	<2.5	>2
	γ				μ	>10	<2.5	>2
Single electron	e	>160	<2.5	>2				
Di-Photon	γ	>60	<2.5	–	γ	>60	<2.5	–
Di-Electron	e	>60	<2.5	>2	e	>60	<2.5	>2
Di-Electron/Photon	e	>60	<2.5	>2	γ	>60	<2.5	–

the full reconstruction chain, this time including **LRT** during track reconstruction, and is saved in the same `xAOD` format as the standard **ATLAS** datasets, now called `DAOD_RPVLL`. These datasets are further processed into `DAOD_SUSY15` derivations, which reduce the size of the dataset by making additional selections on the physics objects which now have the special reconstruction. Finally, the `DAOD_SUSY15` derivations are processed by analysis-specific code into flat n-tuples that can be easily used for studies and background estimates.

6.2 Monte Carlo

Monte Carlo simulation (**MC**) allows analyzers to understand how a given physical process would appear in the **ATLAS** detector. Markov Chain Monte Carlo algorithms are used to randomly sample from the probability distributions from the matrix element calculations and generate many statistically independent events. This results in kinematic distributions that can be compared to data. For this analysis, Monte Carlo samples are generated for various masses and lifetimes of theoretical $\tilde{\ell}$, so that event selection can be optimized for the widest sensitivity. **MC** generation is done by breaking up the process of particle production, decay, and interaction with the detector into many steps.

First, the physical process at the collision is modeled. The center of mass energy must be determined. As discussed in Chap. 3, this is not a known quality and is modeled by **PDF**s. There are many different choices of **PDF**, in this analysis NNPDF23LO is used [1]. Then, the *generator* models the production of the given physical process by calculating its matrix elements. Generally, thought not in this analysis, the generator step models the full process from hard scatter to final state particles including all of their kinematic properties. For this analysis, we use MadGraph5_aMC@NLO [2] to produce $pp \rightarrow \tilde{\ell}\tilde{\ell}$ events with up to two additional radiated partons, using a perturbative **QCD** calculation at leading order accuracy. Then these partons are hadronized and showered into jets using Pythia8.230 using the A14 tuning of the parton showering, hadronization, and modeling of the underlying event [3]. Finally, simulated pileup collisions are overlaid onto the event to mimic the actual conditions at the **LHC**. At the end of this stage, particles are referred to as *truth-level* and are stored in the final **MC** datasets for later study.

Next, the full event is propagated through a simulation of the detector created in GEANT4. Each particle created in the previous steps passes through the detector and all of its magnetic fields and support structures in order to accurately simulate the particle's trajectory, its detector signatures, and other interactions with material. At this stage, the $\tilde{\ell}$ is decayed during this stage and its decay products are then propagated through the simulated detector. GEANT4 does not preserve the information about the chirality of the $\tilde{\tau}$ for the τ decay

Finally, the trajectories of the particles is *digitized*, to emulate the readout of the actual detector, and reconstruction is run on the simulated detector signals in the same way it is run for real data. At the end of this stage, particles are referred to as *reconstruction-level*. Monte Carlo is not required to undergo the same filtering as data, but is processed with **LRT** and made into DAOD_SUSY15 and analysis-specific n-tuples.

The importance of Monte Carlo simulations to an analyzer's ability to search for new physics and measure the **SM** cannot be overstated and an enormous amount of effort is made to make each step as accurate and precise as possible. However, many corrections must be made to these samples after they are generated and even so discrepancies between data and **MC** are an important uncertainty in this analysis and many others.

An *simplified model* is used involving only a few particles and interactions in a given **SUSY** model. **MC** is generated with different $\tilde{\ell}$ flavors as **NSLP!**s (**NSLP!**s), assuming mass degeneracy of the left- and right-handed chiral states. The left- and right-handed $\tilde{\ell}$ are produced according to their cross sections (roughly 2/3 left-handed and 1/3 right-handed). The mixing of $\tilde{\tau}_1$ and $\tilde{\tau}_2$ is set to be $\sin\theta_{\tilde{\tau}} = 0.95$. The mass of the gravitino is set to 0.1 keV (negligibly small) and all other **SUSY** particles have masses too high to be produced at the **LHC**. **MC** samples are produced for a range of possible mass and lifetime configurations of the $\tilde{\ell}$ for $\tilde{e}$ and $\tilde{\mu}$, masses 50 GeV to 1000 GeV, in steps of 100 GeV (after an initial step from 50 GeV to 100 GeV) with lifetimes $\tau = 0.01$ ns, 0.1 ns, 1 ns, and 10 ns; and for $\tilde{\tau}$ up to 600 GeV and lifetimes up to 1 ns. The cross sections for each of these mass points is shown in Tables 6.3 and 6.4. The *signal yield*, the number of signal events that would be seen in the dataset, as a function of $\tilde{\ell}$ lifetime and mass is shown in Figs. 6.1 and 6.2.

Table 6.3 $\tilde{e}$ and $\tilde{\mu}$ signal samples and cross sections. The uncertainty on the cross section calculation ranges from 2–6% and increases with $\tilde{\ell}$ mass

$\tilde{e}/\tilde{\mu}$ mass [GeV]	$\tilde{e}/\tilde{\mu}$ lifetime [ns]	σ [pb]
50	0.01, 0.1, 1, 10	10.736
100	0.01, 0.1, 1, 10	0.7314
200	0.01, 0.1, 1, 10	0.06062
300	0.01, 0.1, 1, 10	0.012508
400	0.01, 0.1, 1, 10	0.003718
500	0.01, 0.1, 1, 10	0.0013472
600	0.01, 0.1, 1, 10	0.0005526
700	0.01, 0.1, 1, 10	0.000247
800	0.01, 0.1, 1, 10	0.00011726
900	0.01, 0.1, 1	5.836×10^{-5}
1000	0.1, 1	3.008×10^{-5}

Table 6.4 $\tilde{\tau}$ signal samples and cross sections. The uncertainty on the cross section calculation ranges from 2–6% and increases with $\tilde{\tau}$ mass

Stau mass [GeV]	Stau lifetime [ns]	σ [pb]
50	0.01, 0.1, 1	5.368
100	0.01, 0.1, 1	0.3657
200	0.01, 0.1, 1	0.03031
300	0.01, 0.1, 1	0.006254
400	0.01, 0.1, 1	0.001859
500	0.01, 0.1, 1	0.0006736

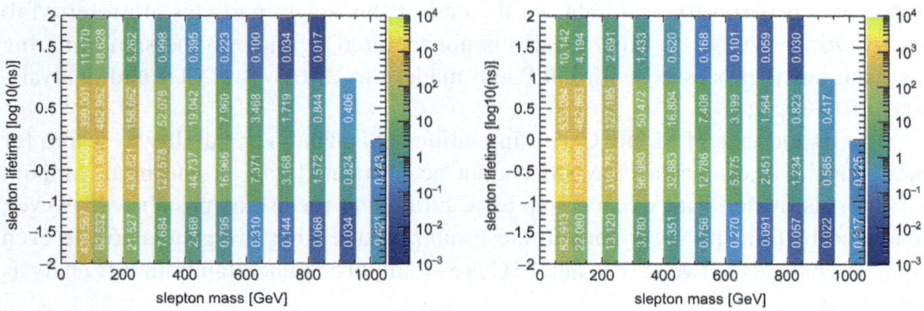

Fig. 6.1 $\tilde{\ell}$ yields for 139 fb^{-1}. Yields are shown in SR-*ee* (left) and SR-$\mu\mu$ (right). The *x*-axis gives the mass of the $\tilde{\ell}$, while the *y*-axis gives their lifetime for three points: 0.01 ns, 0.1 ns, 1 ns, and 10 ns

In addition to the signal **MC** samples, several background **MC** samples are used. However, these have limited use in this analysis because the major backgrounds are not well modeled by **MC**. These include $t\bar{t}$ sample, simulating the production and semi-leptonic decay of two top quarks $pp \to t\bar{t}$; a heavy flavor sample, simulating the production and leptonic decay of two bottom quarks $pp \to BB \to \mu\mu$; and a sample of photons produced by the process $pp \to q\gamma$.

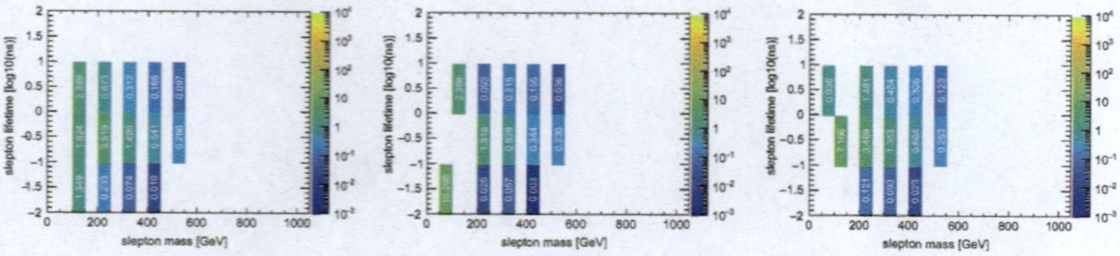

Fig. 6.2 $\tilde{\tau}$ yields for $139\,\text{fb}^{-1}$. Yields are shown in SR-ee (left), SR-$\mu\mu$ (middle), and SR-$e\mu$ (right). The x-axis gives the mass of the status, while the y-axis gives their lifetime for three points: 0.01 ns, 0.1 ns, and 1 ns

References

1. R.D. Ball et al., Parton distributions with LHC data. Nucl. Phys. B **867**, 244 (2013). https://doi.org/10.1016/j.nuclphysb.2012.10.003. arXiv: 1207.1303 [hep-ph]
2. J. Alwall et al., The automated computation of tree-level and next-to-leading order differential cross sections, and their matching to parton shower simulations. JHEP **07**, 079 (2014). https://doi.org/10.1007/JHEP07(2014)079. arXiv: 1405.0301 [hep-ph]
3. ATLAS Collaboration, Summary of ATLAS Pythia 8 tunes. ATL-PHYS-PUB-2012-003 (2012). https://cds.cern.ch/record/1474107

Part III
Long Lived Particle Context

Chapter 7
Searching for Long Lived Particles

A particle is *long lived* if the distance it travels before decaying can be discerned by the detector. This is contrasted with a promptly decaying particle whose decay products point back to the collision point. In **ATLAS**, **LLP** are generally those with a lifetime of more than 1.5 ps (longer than that of a b-hadron) and travel more than 1 mm before decaying.

A **LLP** can decay somewhere inside of the detector or it can pass through the detector without decaying. Note that a detector-stable particle does not necessarily imply a truly stable particle; for example, muons are detector-stable because they never decay within the detector volume. Prompt searches can be used look for neutral stable particles, like dark matter candidates, by calculating **MET** from the prompt particles. **LLP** searches either look directly for the signature of the particle, or look for the products of the metastable decay. This brings technical challenges because standard particle identification algorithms and analysis techniques are not designed for these signatures. It also provides a way to probe processes with extremely small cross sections due to the lack of **SM** processes with the same final state.

This chapter gives an overview of fundamental concepts for **LLP** searches at colliders.

7.1 Lifetime

An individual particle's lifetime cannot be predicted theoretically, but the distribution of lifetimes of a sample of the same particle can be determined from its probability of decay per unit time or *width*, Γ:

$$\Gamma = \frac{1}{2m_X}\int d\Pi_f \left|\mathcal{M}(m_X \to \{p_f\})\right|^2 \tag{7.1}$$

L. Horyn, *A Search for Displaced Leptons in the ATLAS Detector*, Springer Theses,
https://doi.org/10.1007/978-3-030-91672-5_7

where m_X is the mass of the particle, $\mathcal{M}$ is the matrix element for the particle's decay into its decay products $\{p_f\}$ and $d\Pi_f$ is the Lorentz-invariant phase space of the decay. If particle can decay in several different ways, the width of each decay various channel is summed to calculated the particle's *total width*. The branching ratio of a particle decaying via a process i is given as

$$\mathrm{Br}_i = \frac{\Gamma_i}{\Gamma_{\mathrm{total}}} \tag{7.2}$$

Br_i gives the probability that a particle will decay via process i. This is different than the cross section, which measures the probability for a scattering process to result in a given final state. The width is referred to as such because in an experiment with negligible statistical and systematic uncertainties, the width is the full-width-at-half-max of the Gaussian peak of the particle mass spectrum, sketched in Fig. 7.1.

In the sample of particles, the number of particles N that exist at a time t decreases exponentially from the initial number of particles $N(0)$:

$$N(t) = N(0)e^{-\Gamma t} \tag{7.3}$$

A particle's "lifetime", τ, is defined as the time it takes for $\frac{1}{e}$ of the population of particles to decay:

$$\tau = \frac{1}{\Gamma} \tag{7.4}$$

In order for the lifetime of a particle to be long, the width must be small, meaning that the matrix element or allowed phase space must be small. A small matrix element can be caused by a small coupling, or very off-shell intermediate states (like a neutron). A small phase space can be caused by almost-degenerate mass spectra

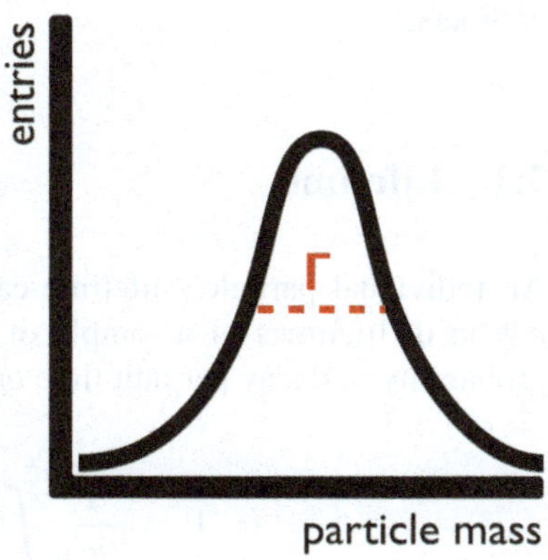

Fig. 7.1 A sketch of a particle mass histogram. The width is shown in red

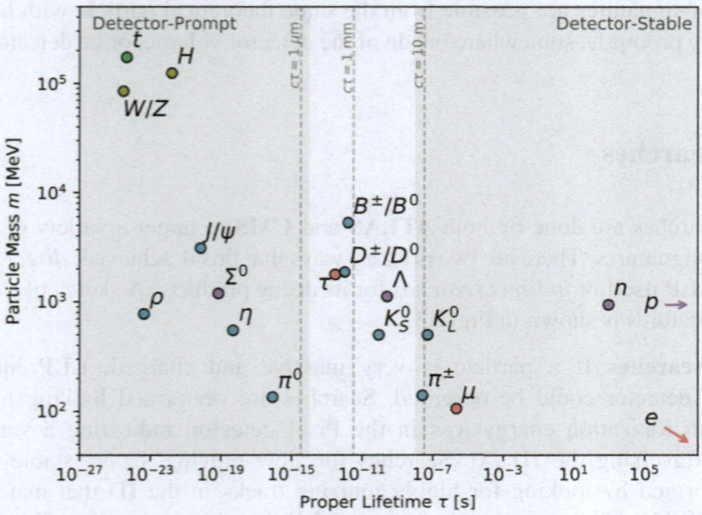

Fig. 7.2 A mass and proper lifetime distribution of particles in the **SM**. There is a wide range of both masses and lifetimes. Shaded regions indicate detector-prompt or detector-stable particles. This assumes that particles traveling at the speed of light $\beta = 1$ [1]

(like a muon). All of these scenarios are common in both the **SM** and many **BSM** theories. The range of lifetimes of selected **SM** particles can be seen in Fig. 7.2 [1].

7.2 Particle Decays

The distance traveled by any relativistic particle can be described by

$$d = \beta\gamma t \tag{7.5}$$

where γ and β have the usual special relativistic definitions: $\gamma = E/m = (1 - \beta^2)^{(-\frac{1}{2})}$, $\beta = v/c = |\vec{p}|/E$, v is the particle's velocity, t is the time during which the particle moves, and c is the speed of light. In the laboratory frame, the distance the particle travels before decaying is given by sending $t \to \tau_0$ in Eq. 7.5, where τ_0 denotes the proper lifetime of the particle measured in its own rest frame.

There is an exponential probability that the particle will decay at a given time (t), given by

$$P(t) = e^{-t/(\gamma\tau)} \tag{7.6}$$

A particle with a large τ has a wider probability distribution and there is a larger range of times in which the particle can decay. Practically, this means that a wide variety of signatures are possible from the same theoretical particle with large τ: it can decay promptly, somewhere inside of the detector volume, or be detector-stable.

7.3 Searches

Many searches are done by both **ATLAS** and **CMS** to target a variety of possible detector signatures. There are two primary ways that this is achieved: *direct searches* for the **LLP** itself or *indirect searches* for its decay products. A sketch of a different **LLP** signatures is shown in Fig. 7.3.

Direct Searches If a particle is very massive and charged, **LLP** interaction with the detector could be observed. Searches are performed looking for tracks have high ionization energy loss in the Pixel detector, indicating a very heavy particle traversing the **ID** [3]. Searches for slow moving, heavy stable particles are performed by looking for highly ionizing tracks in the **ID** that match tracks in the **MS** [4]. Other searches look for a **LLP** that decays into invisible particles by looking for short, disappearing tracks [5, 6]. These signatures generally involve prompt tracks.

Indirect Searches If more than one of the products of the **LLP** decay are visible to the detector, one can look for its decay vertex. The **LLP** will travel some distance from the **PV**, where the particle is produced, and its decay vertex will be

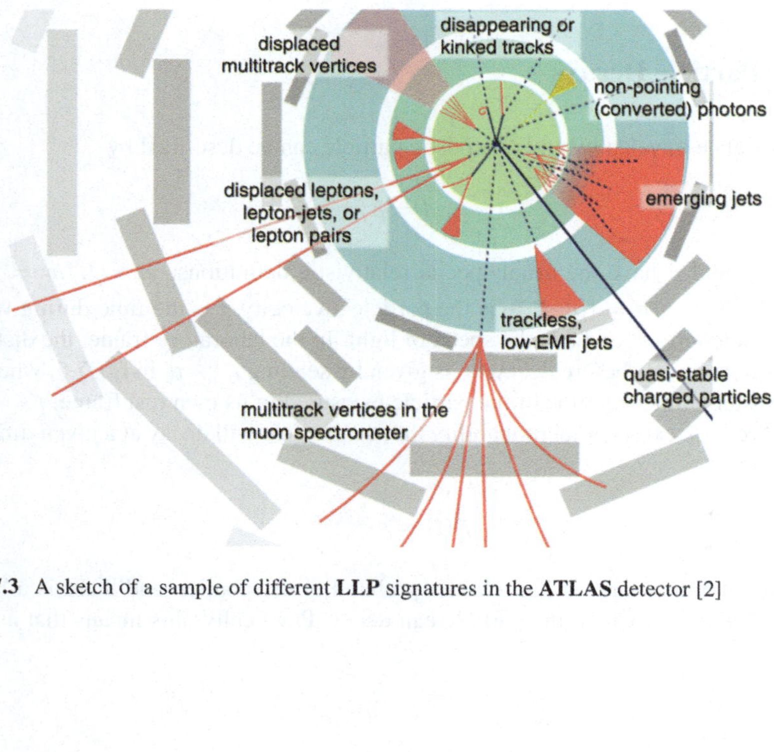

Fig. 7.3 A sketch of a sample of different **LLP** signatures in the **ATLAS** detector [2]

displaced, called a Displaced Vertex (**DV**). In this case, direct information about τ_0 of the particle can be deduced from the difference in distance between the **LLP**'s production and decay vertices. This vertices can appear in either tracking volume of the detector (**ID** or **MS**). There are many searches for **DV**s in **ATLAS**, many in conjunction with other physics objects like leptons, jets, or **MET** depending on the particular theory model being probed [5, 7–11].

One can also look for the decay products of the **LLP** without a **DV** via unconventionally presenting physics objects. Examples include photons that do not point back to the **PV** in the **EM** calorimeter [12], or emerging jets, which become visible midway through their hadronization [13]. In this analysis, the target signature is leptons whose trajectories do not point back to the **PV**.

For indirect searches, it is often convenient to use d_0, the transverse impact parameter, as a measure of particle displacement. The d_0 is a parameter of **ID** track fit, discussed in Sect. 5.1. It does not give direct information about τ_0, but it can be known with high precision. The distance the particle traveled from the collision, R, would give direct information about the proper lifetime of the particle, but the detector can only provide the radius of the first **ID** hit in the track. The distance between silicon **ID** layers at minimum 30 mm and any requirements on R are very sensitive to transient detector failures. d_0, however, has a $\mathcal{O}(10\,\mu\text{m})$ resolution, shown in Fig. 5.3. The d_0 of the decay products of a **LLP** with a longer lifetime have wider d_0 distributions as seen in Fig. 7.4. The d_0 of the decay products is also correlated with the momentum of the **LLP**, shown in Fig. 7.5. This relationship induces a correlation between the p_T and d_0 of the decay products.

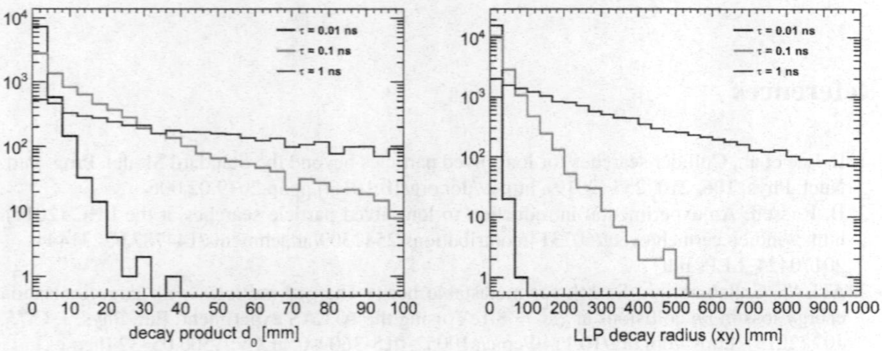

Fig. 7.4 d_0 (left) and transverse decay radius $(x - y)$ of **LLPs** with different τ. The left plot demonstrates that while d_0 does not directly give information about τ, decay products of particles with longer lifetimes have wider d_0 distributions. These plots are made at truth level

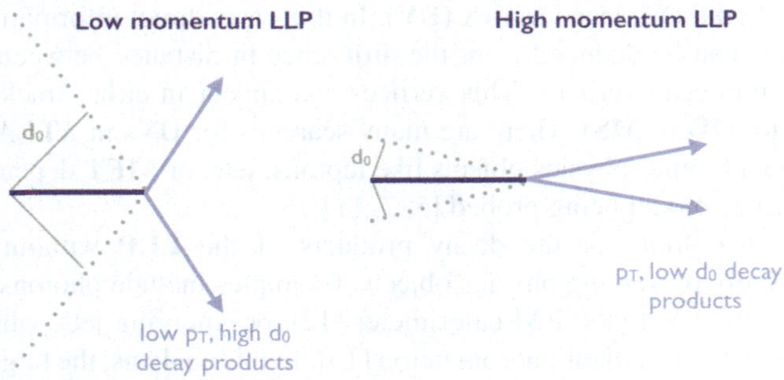

Fig. 7.5 A sketch of the relationship between the momentum of the parent **LLP** and the d_0 of the daughter particles

7.3.1 *Common Strategies*

Due to the technical challenges associated with **LLP** searches, analyzers want their result to be applicable to as many potential **BSM** particles as possible and they aim to make their selections as *model-independent* as possible. This means that, while there is usually a model to which a search brings unique sensitivity, the search not over-optimized for this model. For example, in this analysis, the events are only required to have two displaced leptons. The slepton decays in the **GMSB SUSY** model include **MET** from the gravitino, but no requirement is made to ensure the result is sensitive to other possible models that might not include **MET**.

References

1. L. Lee et al., Collider searches for long-lived particles beyond the Standard Model. Prog. Part. Nucl. Phys. **106**, 210–255 (2019). https://doi.org/10.1016/j.ppnp.2019.02.006
2. H. Russell, An experimental introduction to long-lived particle searches at the LHC (2017). https://indico.cern.ch/event/607314/contributions/2542309/attachments/1447873/2231444/20170424_LLPs.pdf
3. ATLAS Collaboration, Search for metastable heavy charged particles with large ionisation energy loss in *pp* collisions at $\sqrt{s} = 8$ TeV using the ATLAS experiment. Eur. Phys. J. C **75**, 407 (2015). https://doi.org/10.1140/epjc/s10052-015-3609-0. arXiv: 1506.05332 [hep-ex]
4. ATLAS Collaboration, Search for heavy long-lived charged R-hadrons with the ATLAS detector in 3.2 fb^{-1} of proton–proton collision data at $\sqrt{s} = 13$ TeV. Phys. Lett. B **760**, 647 (2016). https://doi.org/10.1016/j.physletb.2016.07.042. arXiv: 1606.05129 [hep-ex]
5. ATLAS Collaboration, Search for long-lived charginos based on a disappearing-track signature in *pp* collisions at $\sqrt{s} = 13$ TeV with the ATLAS detector. JHEP **06**, 022 (2018). https://doi.org/10.1007/JHEP06(2018)022. arXiv: 1712.02118 [hep-ex]
6. CMS Collaboration, Search for disappearing tracks as a signature of new long-lived particles in proton–proton collisions at $\sqrt{s} = 13$ TeV. JHEP **08**, 016 (2018). https://doi.org/10.1007/JHEP08(2018)016. arXiv: 1804.07321 [hep-ex]

7. ATLAS Collaboration, Search for massive, long-lived particles using multitrack displaced vertices or displaced lepton pairs in *pp* collisions at $\sqrt{s} = 8$ TeV with the ATLAS detector. Phys. Rev. D **92**, 072004 (2015). https://doi.org/10.1103/PhysRevD.92.072004. arXiv: 1504.05162 [hep-ex]
8. ATLAS Collaboration, Search for long-lived, massive particles in events with a displaced vertex and a muon with large impact parameter in *pp* collisions at $\sqrt{s} = 13$ TeV with the ATLAS detector. Phys. Rev. D **102**(3), 032006 (2020). https://doi.org/10.1103/PhysRevD.102.032006. arXiv: 2003.11956 [hep-ex]
9. CMS Collaboration, Search for long-lived particles decaying into displaced jets in proton–proton collisions at $\sqrt{s} = 13$ TeV. Phys. Rev. D **99**, 032011 (2019). https://doi.org/10.1103/PhysRevD.99.032011. arXiv: 1811.07991 [hep-ex]
10. CMS Collaboration, Search for long-lived particles with displaced vertices in multijet events in proton–proton collisions at $\sqrt{s} = 13$ TeV. Phys. Rev. D **98**, 092011 (2018). https://doi.org/10.1103/PhysRevD.98.092011. arXiv: 1808.03078 [hep-ex]
11. CMS Collaboration, Search for R-parity violating supersymmetry with displaced vertices in proton–proton collisions at $\sqrt{s} = 8$ TeV. Phys. Rev. D **95**, 012009 (2017). https://doi.org/10.1103/PhysRevD.95.012009. arXiv: 1610.05133 [hep-ex]
12. ATLAS Collaboration, Search for nonpointing and delayed photons in the diphoton and missing transverse momentum final state in 8 TeV *pp* collisions at the LHC using the ATLAS detector. Phys. Rev. D **90**, 112005 (2014). https://doi.org/10.1103/PhysRevD.90.112005. arXiv: 1409.5542 [hep-ex]
13. CMS Collaboration, Search for new particles decaying to a jet and an emerging jet. JHEP **02**, 179 (2019). https://doi.org/10.1007/JHEP02(2019)179. arXiv: 1810.10069 [hep-ex]

Chapter 8
Experimental Context

As discussed in Sect. 2.2.2, the lack of coverage of the displaced dilepton signature leaves a sensitivity gap in **LHC**-accessible **SUSY**. All other searches for signatures that include displaced leptons also required a secondary vertex and are not sensitive to $\tilde{\ell}$ production. The two ℓ from pair-produced $\tilde{\ell}$ do not originate from the same decay so they are not connected by a vertex. Furthermore, since the $\tilde{G}$ is uncharged it leaves no track so there is no secondary vertex in the event.

The lifetime of the $\tilde{\ell}$ is related to the **SUSY** breaking scale (see Eq. 2.5) which is not known, thus a suite of searches must be done to study the range of detector signatures corresponding to the range of possible lifetimes. This chapter will discuss the two previous searches for the displaced leptons signature, as well as the other searches that provide sensitivity to long lived $\tilde{\ell}$ with different lifetimes. The different track signatures discussed in this chapter are shown in Fig. 8.1.

8.1 Previous Searches for Displaced Leptons

The most stringent limits on long lived sleptons are the result of a searches from **LEP** [1–5]. A search for different flavor ($e\mu$) displaced leptons was performed by **CMS**, but did not target long lived sleptons and did not surpass the limit set at LEP [6]. There are no **ATLAS** searches that would select events with displaced leptons and no secondary vertex.

8.1.1 LEP

The **LEP** experiments searched for long-lived $\tilde{\ell}$ signatures from the process $e^+e^- \rightarrow \tilde{\ell}_R^+\tilde{\ell}_R^-$. A combination of results from the LEP experiments (OPAL, L3,

L. Horyn, *A Search for Displaced Leptons in the ATLAS Detector*, Springer Theses,
https://doi.org/10.1007/978-3-030-91672-5_8

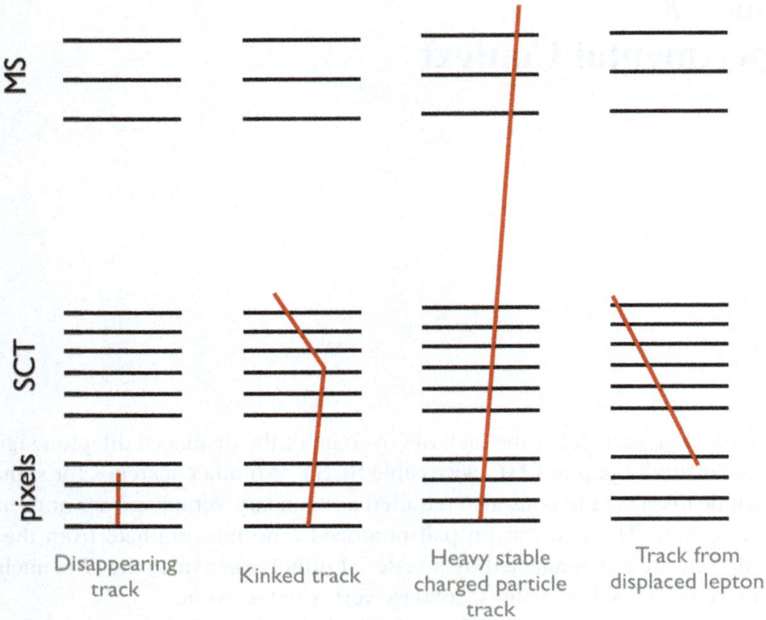

Fig. 8.1 A sketch of the various track signatures in a simplified **ATLAS** detector that give sensitivity to displaced leptons. In reality, the tracks are curved by the magnetic fields in the **ID** and **MS**

ALEPH, and DELPHI) excluded right-handed $\tilde{\mu}$ and $\tilde{e}$ of all lifetimes with masses less than 96.3 GeV and 65.8 GeV, respectively. The OPAL experiment alone set the best limits on all lifetimes of $\tilde{\tau}_1$, a mixture of left- and right-handed states, with masses less than 87.6 GeV. The OPAL search will be discussed in detail.

The OPAL experiment at **LEP** set limits on long-lived $\tilde{\ell}$ signatures and set limits on $\tilde{\ell}$ lifetimes between 10^{-6} s (detector-stable) and 10^{-12} s (prompt signatures). The detector conditions of the *ee* collider are much less busy than those of the *pp* collider—the extra particles and tracks created by low energy **QCD** interactions between other partons in the proton or other protons in the bunch do not exist, simplifying the task of track reconstruction. The search was done with 600 pb^{-1} of data with $\sqrt{s} = 189 - 209$ GeV between the years 1998 and 2000.

The portion of the analysis that targets displaced leptons has a relatively simple event selection. Events are required to have no prompt tracks and at least two tracks with opposite charges and $d_0 > 5$ mm corresponding to the displaced leptons. To reduce backgrounds from cosmic-ray muons, a time-of-flight detector was used. To reduce beam-induced backgrounds, both tracks were required to have $|z_0| < 400$ mm. To reduce background from Bhabha scattering ($e^+e^- \rightarrow e^+e^-$ by exchanging a γ), the tracks were required to not be back-to-back. To reduce the background from di-photon events, tracks must have $pt > 10$ GeV and the invariant mass of the two tracks must be greater than 5 GeV. 0.13 ± 0.76 events were expected and zero were observed. To specifically probe the $\tilde{\tau}$ decay, the track p_T and invariant mass requirements were loosened to 5 GeV and 3 GeV, respectively. Here, 3.79 ± 1.71 events were expected and 3 observed.

Additionally, this search targeted $\tilde{\ell}$ with shorter lifetimes by looking for events with at least two leptons and **MET**. Longer lifetimes were probed by looking for kinked tracks (see Fig. 8.1) where the tracks of the $\tilde{\ell}$ and ℓ are both observed. Signatures of $\tilde{\ell}$ with even longer lifetimes and escape the detector were targeted by looking for tracks with high ionization energy loss. No excess of events was observed in any of the signatures, and the limits set are shown in Fig. 8.2.

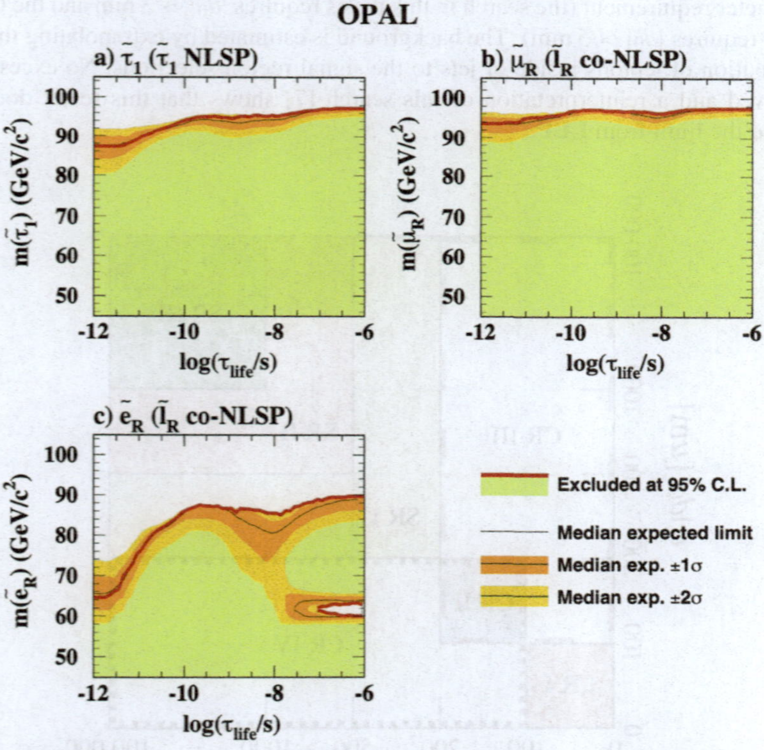

Fig. 8.2 Limits on $\tilde{\ell}$ production set by the OPAL experiment for the $\tilde{\tau}$ **NLSP** scenario (top left) and $\tilde{\mu}$ (top right) and $\tilde{e}$ (bottom left) in the co-**NLSP** scenario [1]

8.1.2 CMS

The only search for vertexless displaced leptons at the **LHC** is from the **CMS** experiment, and it did not target $\tilde{\ell}$ decays. The target model used for this search is the RPV $\tilde{t}$ decay described in Sect. 2.2.3. The search is for differently flavored leptons ($e\mu$) only. 2.6 fb^{-1} of $\sqrt{s}$ = 13 TeV data taken during 2015 was used.

The events are collected using a trigger that selects events with an electron and a muon and are then categorized into three signal regions, defined by the $|d_0|$ of the leptons shown in Fig. 8.3. The leptons must have $p_T > 25$ GeV, $|d_0| > 0.1$ mm, opposite electric charges, and be separated ($\Delta R > 0.4$). Dominant backgrounds are from decays of heavy flavor hadrons (like b-hadrons) due to the small impact parameter requirement (the search in this thesis requires $|d_0| > 3$ mm and the OPAL result requires $|d_0| > 5$ mm). The background is estimated by extrapolating the $|d_0|$ distribution of leptons inside of jets to the signal region selections. No excess was observed and a reinterpretation of this search [7] shows that this result does not exceed the limit from **LEP**.

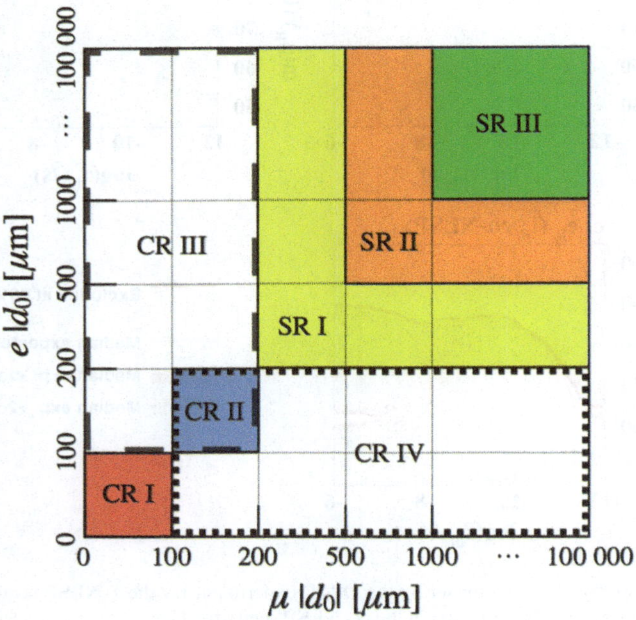

Fig. 8.3 Signal and control regions used by CMS in Ref. [6]. The signal regions have significantly lower $|d_0|$ (>0.2 mm) than in this thesis search (>3 mm)

8.2 Other Signatures with Sensitivity to Long Lived Sleptons

There are no constraints on the possible lifetime of a $\tilde{\ell}$ in a **GMSB** model, so various searches must be done together to cover the whole phase space. The displaced lepton signature covers an intermediate lifetime range. Searches for longer lifetimes look directly for the signature of the $\tilde{\ell}$ instead of its decay products. $\tilde{\ell}$ that decay in the tracker are covered by disappearing track searches [8–11] and those which are detector stable are probed by searches for tracks consistent with those from a heavy stable charged particle [9, 12, 13]. These signatures have been covered by both **ATLAS** and **CMS**. While they did not directly target long lived $\tilde{\ell}$, the reinterpretation done by Ref. [7] shows that they improve significantly over the OPAL result. All constraints on $\tilde{\ell}$ production as determined by Ref. [7] are shown in Fig. 8.4.

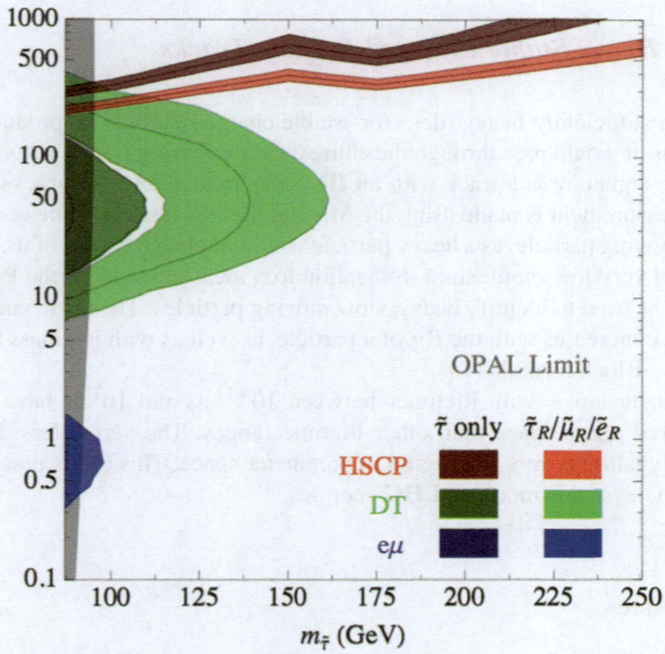

Fig. 8.4 Constraints on direct production of long lived $\tilde{\ell}$. $\tilde{\tau}$ and co-NLSP scenarios are shown in darker and brighter colors, respectively. The OPAL result discussed in Sect. 8.1.1 is shown in gray and **CMS** result discussed in Sect. 8.1.2 is shown in blue. Constraints from disappearing track signatures (DT) discussed in Sect. 8.2.1 is shown in green, and constraints from heavy stable charged particle signatures (HSCP) discussed in Sect. 8.2.2 is shown in red. The **LHC** searches are from Run 1, with $\sqrt{s} = 8$ TeV. The search in this thesis covers the $c\tau$ range 0.01–100 cm with masses up to 800 GeV for co-NLSP scenarios [7]

8.2.1 Disappearing Tracks

A *disappearing track* is a short track measured in the Inner Detector (**ID**). This track has hits in the innermost **ID** layers and no hits along the rest of its trajectory In reality, the signature from a $\tilde{\ell}$ decay is a kinked track, as targeted by OPAL. However, looking for kinked tracks is extremely difficult in the **LHC** environment as kink be at any angle and any point along the track.

Both **ATLAS** and **CMS** look for a high p_T track with no hits in the outer layers of the tracking detector. These events are triggered based on their large **MET** resulting from gravitino. Events are also required to have at least one high p_T jet ($p_T >$ 90 GeV) with other jets well separated from each other and the **MET**. They must also have no muon or electron candidates. This signature has sensitivity to $\tilde{\tau}$ with hadronic τ decays.

8.2.2 Heavy Stable Charged Particle Tracks

Should a sufficiently heavy (detector-)stable charged particle be produced in **LHC** collisions, it would pass through the entire detector leaving tracks in both the **ID** and **MS**. The signature is a track with an **ID** tracks matched to **MS** tracks. A time-of-flight measurement is made using the **MS** and the **MS** track must be consistent with a slow moving particle, as a heavy particle would likely have most of its energy in its mass and very low momentum. Ionization loss measurements in the Pixel detector can also be used to identify heavy, slow moving particles. The mean energy loss per unit space increases with the $\beta\gamma$ of a particle, as well as with it's mass according to the Bethe-Bloch formula [14].

In conclusion, $\tilde{\ell}$ with lifetimes between 10^{-11} ns and 10^{1} ns have much more unexplored phase space than other lifetime ranges. The vertex-less displaced di-lepton signature covers this region of parameter space. This thesis presents the first search to target this model at **LHC** energies.

References

1. The OPAL Collaboration, Searches for Gauge-Mediated Supersymmetry Breaking topologies in e^+e^- collisions at centre-of-mass energies up to $\sqrt{s} = 209$ GeV. Eur. Phys. J. C **46**(2), 307–341 (2006). https://doi.org/10.1140/epjc/s2006-02524-8
2. J. Abdallah et al., Searches for supersymmetric particles in e^+e^- collisions up to 208 GeV and interpretation of the results within the MSSM (2003)
3. J. Abdallah et al., Search for supersymmetric particles in light gravitino scenarios and sleptons NLSP. Eur. Phys. J. C **27**, 153–172 (2003). https://doi.org/10.1140/epjc/s2002-01112-4. arXiv: hep-ex/0303025
4. ALEPH Collaboration, Search for gauge mediated SUSY breaking topologies in e^+e^- collisions at center-of-mass energies up to 209-GeV. Eur. Phys. J. C **25**, 339–351 (2002). https://doi.org/10.1007/s10052-002-1005-z. arXiv: hep-ex/0203024

5. ALEPH, DELPHI, L3, OPAL Experiments. Combined LEP Selectron/Smuon/Stau Results, 183-208 GeV. LEPSUSYWG/04-01.1 (2004). http://lepsusy.web.cern.ch/lepsusy/www/sleptons_summer04/slep_final.html
6. Search for displaced leptons in the e-mu channel. Technical report. CMS-PAS-EXO-16-022. Geneva: CERN (2016). https://cds.cern.ch/record/2205146
7. J. Evans, J. Shelton, Long-lived staus and displaced leptons at the LHC. J. High Energy Phys. **2016** (2016). https://doi.org/10.1007/JHEP04(2016)056
8. ATLAS Collaboration, Search for long-lived charginos based on a disappearing-track signature in *pp* collisions at $\sqrt{s} = 13$ TeV with the ATLAS detector. JHEP **06**, 022 (2018). https://doi.org/10.1007/JHEP06(2018)022. arXiv: 1712.02118 [hep-ex]
9. CMS Collaboration, Search for disappearing tracks as a signature of new long-lived particles in proton–proton collisions at $\sqrt{s} = 13$ TeV. JHEP **08**, 016 (2018). https://doi.org/10.1007/JHEP08(2018)016. arXiv: 1804.07321 [hep-ex]
10. ATLAS Collaboration, Search for anomaly-mediated supersymmetry breaking with the ATLAS detector based on a disappearing-track signature in *pp* collisions at $\sqrt{s} = 7$ TeV. Eur. Phys. J. C **72**, 1993 (2012). https://doi.org/10.1140/epjc/s10052-012-1993-2. arXiv: 1202.4847 [hep-ex]
11. CMS Collaboration, Search for disappearing tracks in proton–proton collisions at $\sqrt{s} = 8$ TeV. JHEP **01**, 096 (2015). https://doi.org/10.1007/JHEP01(2015)096. arXiv: 1411.6006 [hep-ex]
12. ATLAS Collaboration, Search for heavy long-lived charged R-hadrons with the ATLAS detector in 3.2 fb^{-1} of proton–proton collision data at $\sqrt{s} = 13$ TeV. Phys. Lett. B **760**, 647 (2016). https://doi.org/10.1016/j.physletb.2016.07.042. arXiv: 1606.05129 [hep-ex]
13. ATLAS Collaboration, Search for heavy long-lived charged particles with the ATLAS detector in *pp* collisions at $\sqrt{s} = 7$ TeV. Phys. Lett. B **703**, 428 (2011). https://doi.org/10.1016/j.physletb.2011.08.042. arXiv: 1106.4495 [hep-ex]
14. M. Tanabashi et al., Review of particle physics. Phys. Rev. D **98**(3), 030001 (2018). https://doi.org/10.1103/PhysRevD.98.030001

Part IV
Search for Displaced Leptons

Chapter 9
Event Selection

9.1 Analysis Strategy

This analysis aims to select events that would have small cross sections producing only a handful of events from the entire 139 fb^{-1} collected by **ATLAS**. The signature of two displaced leptons, is rather simple in principle, but this analysis had not been done before in the conditions of the **LHC**, so extensive optimization of the signal regions and lepton selection criteria needed to be done. A cut in the event selection should decrease or eliminate some background, but have little to no impact on the signal. For example, quality criteria ensure that leptons are well measured in the detector, and leptons from slepton decays will pass these quality criteria, but fake leptons from reconstruction failures generally will not. Every effort was made to remain as signal-agnostic as possible, so cuts on the specific topology of the **GMSB SUSY** decay were avoided.

This analysis is split into three signal regions, defined by the flavor of the two highest p_T leptons in the event: SR-*ee* with two electrons, SR-$\mu\mu$ with two muons, and SR-$e\mu$ with an electron and a muon. No requirement is placed on the charge of the leptons. There are two definitions of leptons used: *baseline* and *signal*. Baseline leptons are required to pass the reconstruction and identification criteria described in Sects. 5.4.2 and 5.3.2 and have $p_T > 50$ GeV and $|d_0| > 2$ mm. Signal muons are required to further pass bespoke quality requirements, an isolation cut, and have $p_T > 65$ GeV and $|d_0| > 3$ mm. Displacement-independent quality variables, described in the rest of this chapter, were defined specifically for this analysis as many standard quality criteria place requirements on the $|d_0|$, $|z_0|$, or the number of hits in the Pixel detector, all of which would limit the ability to target displaced leptons.

L. Horyn, *A Search for Displaced Leptons in the ATLAS Detector*, Springer Theses,
https://doi.org/10.1007/978-3-030-91672-5_9

9.1.1 Background Overview

The two signal leptons are required to have high p_T and high $|d_0|$, higher than that expected of any leptons resulting from **SM** processes. The remaining backgrounds come from failures of reconstruction algorithms, extreme tails of decays of Heavy Flavor (**HF**) physics (b-jets or τs), and muons from cosmic rays. None of these background sources are well modeled in **MC**, so data-driven methods must be used. This brings additional challenges from statistical limitations and care must be taken to avoid unblinding the signal regions.

Fake leptons result from the mis-association of a track to a calorimeter or **MS** signature, creating a reconstructed lepton that does not correspond to a real particle that passed through the detector. This is particularly likely due to use of **LRT**. The extended tracking algorithm loosens several requirements that enable high-$|d_0|$ tracking, but also introduces many *fake tracks*, combinations of hits that do not correspond to the trajectory of a particle in the detector. When matched to **MS** or calorimeter signatures, high-$|d_0|$ leptons can be created and form a background to this search. This is particularly likely to occur during electron reconstruction, where the track can be associated to real calorimeter signatures from photons or converted photons.

The background to SR-$\mu\mu$ is dominated by cosmic ray muons. The **ATLAS** detector is far underground partially to avoid the many particles from cosmic rays passing through the detector. However, there is a service shaft above the detector and muons can make it through the earth above the detector, so muons from cosmic rays are constantly passing through the detector. If a muon from a cosmic ray passes through the detector and is coincident with a bunch crossing, the event could be triggered and that muon reconstructed. Since it can pass at any point with respect to the collision, it can be reconstructed as one or two muons with high $|d_0|$, exactly mimicking the targeted signature.

Both electrons and muons suffer potential backgrounds from **HF** decays, predominately from the decays of b-hadrons. b-quarks are very commonly produced in **LHC** collisions and they immediately hadronize into b-hadrons, which have a lifetime of 1.5 ps and whose decays include a lepton 11% of the time [1]. This is sufficiently long that the b-hadron travels far enough before it decays such that a secondary vertex can be identified. Generally, the tracks have low p_T and maximum $|d_0|$ of about 1.5 mm, and the 3 mm cut on signal leptons is designed to minimize this effect. However, it is possible that extreme tails of this distribution are not well modeled in **MC** and in an analysis where 0 background events are expected, understanding each possible source is paramount.

9.2 Event Requirements

Events must pass a trigger in order to be recorded by the **ATLAS** detector. Three different triggers are used in this analysis and the data separated into three orthogonal regions based on the topology of the event, described in Table 6.1. The trigger is required to pass in the appropriate region for the event to be selected.

Each event is required to pass a standard set of **ATLAS** event quality preselection criteria. Specifically, these include detector error flags which reject events with noise bursts or data corruption, or events in periods where any sub-detector was operating suboptimally. Events are required to have at least one **PV** with $|z| < 200$ mm.

From events that pass the previous two requirements, events are sorted into the three signal regions based on the highest p_{T} baseline leptons in the event. Leptons in signal events are required to be well separated with $\Delta R_{\ell\ell} > 0.2$. This eliminates background from lepton pairs that could be created from an interaction with the material of the detector. Finally, signal events are required to have zero cosmic muons. The cosmic tag and the associated background will be discussed in Sect. 10.3.1. The analysis was optimized and the backgrounds estimated while keeping the Signal Regions (**SRs**) blinded, so Control Regions (**CRs**) (where backgrounds are estimated) and Validation Regions (**VRs**) (where additional studies and validation are done) are defined with different numbers of leptons and cosmic tags. All **CRs** and **VRs** are designed to be dominated by backgrounds and have very few signal events in them. A full list of all signal, control, and validation regions can be seen in Table 9.1.

9.3 Electrons

Quality cuts specifically designed for this analysis are introduced to eliminate fake electrons reconstructed from the incorrect association of tracks and calorimeter clusters. The most important of these cuts is

$$\Delta p_{\mathrm{T}}/p_{\mathrm{T}} \equiv \frac{p_{T,\mathrm{track}} - p_{T,e}}{p_{T,e}} \tag{9.1}$$

which measures the degree of consistency between the p_{T} of the electron, which is dominated by the calorimeter measurement, and the p_{T} of the track. Fake tracks tend to be low p_{T}, resulting in fake electrons with tracks with less than half of the p_{T} of the electron, $\Delta p_{\mathrm{T}}/p_{\mathrm{T}} < -0.5$.

Additionally, quality requirements are imposed on the **ID** tracks. They are required to have $\chi^2_{\mathrm{ID}} < 2$[1] and no more than 1 missing hit after the first hit in the track. These cuts remove poor hit combinations, short tracks, or tracks missing

[1] In this thesis, χ^2 implies χ^2 divided by the number of degrees of freedom in the fit.

Table 9.1 Summary of signal, control and validation regions used in the analysis. All regions are defined exclusively by their reconstructed leptons. In the table, all leptons should be assumed to be signal leptons except for their noted deviation from the signal requirements. All requirements are placed on the two leading leptons, additional leptons are allowed in the event but no selections are made on them. In each region, the appropriate trigger selection is made. In region names, a capital M denotes a cosmic-tagged muon, and for requirements on numbers of cosmics, all leptons in the event are considered. An * on the number of cosmic tags denotes the number of muons tagged or untagged by the narrow tag, not the nominal tag used in the signal regions

Purpose	Name	# of Leptons	# of Cos.Tags	Additional requirements
Signal regions	SR-ee	$\geq 2\,e$	0	
	SR-$\mu\mu$	$\geq 2\,\mu$	0	
	SR-$e\mu$	$\geq 1\,e$, $\geq 1\,\mu$	0	
Control regions				
Fake estimation	CR-ee-fake	$\geq 2\,e$	0	≥ 2 loosened electrons, not in SR-ee
	CR-$e\mu$-fake	$\geq 1\,e$, $\geq 1\,\mu$	0	≥ 2 loosened leptons, not in SR-$e\mu$
Heavy flavor estimation	CR-$\mu\mu$-hf	$\geq 2\,\mu$	0	≥ 1 anti-isolated ℓ, loosened p_T and d_0
Cosmic estimation	CR-M_{full}	$\geq 1\,\mu$	≥ 1	Includes muons failing N_ϕ and N_{prec} cuts
	CR-$\mu\mu$-topbad	$\geq 2\,\mu$	0	One signal and one loosened muon
Validation regions				
Cut evaluation	VR-M_{narrow}	$\geq 1\,\mu$	≥ 1	Using narrow cosmic tag
	VR-e	$1\,e$, $0\,\mu$	0	Electron is baseline
	VR-μ	$0\,e$, $1\,\mu$	0	Muon is baseline
Fake validation	VR-ee-fake	$\geq 2\,e$	0	Inverted $\Delta p_T/p_T$ selection
	VR-ee-fake-hf	$\geq 2\,e$	0	≥ 1 anti-isolated ℓ, loosened quality
	VR-$e\mu$-fake	$\geq 1\,e$, $\geq 1\,\mu$	0	$1\,e$ fails $\Delta p_T/p_T$, $1\,\mu$ fails χ^2_{CB}
	VR-$e\mu$-fake-hf	$\geq 1\,e$, $\geq 1\,\mu$	0	$1\,e$ fails $\Delta p_T/p_T$, no isolation req., loosened quality
Cosmic validation	VR-μM_{full}	$\geq 2\,\mu$	1	
	VR-μ-narrow	$\geq 1\,\mu$	0*	Using narrow cosmic tag
	VR-μM_{narrow}	$\geq 2\,\mu$	1*	Using narrow cosmic tag

Table 9.2 Overview of electron signal selections

Electron selections	
p_{T}	>65 GeV
$\|d_0\|$	>3 mm
$\|\eta\|$	<2.47
Isolation	`FCTight`
$\Delta p_{\mathrm{T}}/p_{\mathrm{T}}$	≥ -0.5
χ^2_{ID}	<2
N_{miss}	≤ 1

measurements. Additionally, signal electrons are required to be isolated and pass the `FCTight` isolation definition described in Sect. 5.6 (Table 9.2).

9.4 Muons

Signal muon requirements are similar to those placed on signal electrons. The muon is required to have $\chi^2_{\mathrm{CB}} < 3$ in order to ensure a good combination between **ID** track and **MS** track, similar to the $\Delta p_{\mathrm{T}}/p_{\mathrm{T}}$ cut on electrons. Additionally, signal muons are required to pass the same **ID** track quality requirements as electrons as well as **MS** track quality requirements. Requiring $N_{\mathrm{prec}} \geq 3$ (number of **MDT** layers with at least 3 hits) and $N_{\phi} \geq 1$ (number of **RPC** hits) ensure that the **MS** track is well measured in both its η and ϕ coordinates. Signal muons required to pass the `FCTight` isolation requirement described in Sect. 5.6.

Furthermore, muons are required to have $t_0^{\mathrm{avg}} < 30$. t_0^{avg} is the average t_0 calculated from the **MS** segments associated to the muon. The measurement of t_0 is described in Sect. 4.4.1.1. If the fit fails, the segment is assigned $t_0 = 0$, and this value does not enter the t_0^{avg} calculation, but muons that have all segments with $t_0 = 0$ are not rejected. This cut is designed to better control the background from cosmic muons, which have a much wider spread in t_0^{avg} than collision muons (Table 9.3).

9.5 Acceptance and Efficiency

Acceptance is defined as the fraction of events that could enter the **SR** based on their kinematics, and efficiency is fraction of the accepted events that get correctly identified. The acceptance and efficiency in this analysis are both rather low. The exponential nature of the p_{T} and $|d_0|$ distributions of the $\tilde{\ell}$ decays, many daughter leptons have low p_{T} and low $|d_0|$ and do not pass the signal kinematic selections and are not accepted. This particularly effects low mass, low lifetime $\tilde{\ell}$; the p_{T} cut has an even stronger impact on leptons from $\tilde{\tau}$ decays, which must decay through a standard model τ. Conversely, the degradation of the **LRT** efficiency at high $|d_0|$

Table 9.3 Overview of muon signal selections

Muon selections	
p_{T}	>65 GeV
$\lvert d_0 \rvert$	>3 mm
$\lvert \eta \rvert$	<2.5
Isolation	`FCTight`
N_{prec}	≥3
χ^2_{CB}	<3
N_{ϕ}	≥1
χ^2_{ID}	<2
N_{miss}	≤1
$\lvert t_0^{\mathrm{avg}} \rvert$	<30
Pass Cosmic Veto	True

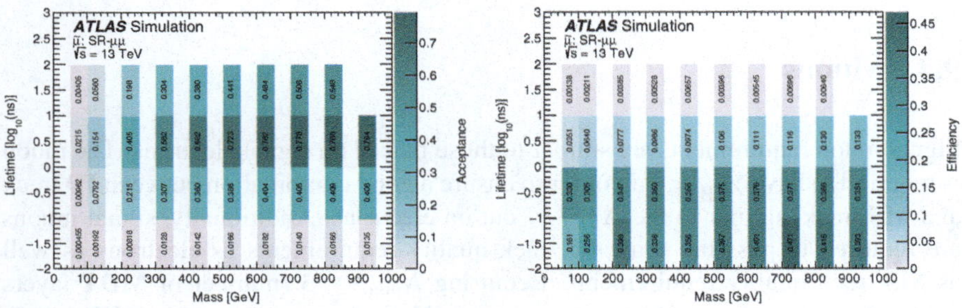

Fig. 9.1 Acceptance (left) and efficiency for $\tilde{\mu}$ decaying to muons in SR-$\mu\mu$. The x-axis shows the possible masses of the $\tilde{\mu}$ and the y-axis its possible lifetime

means that leptons with high $|d_0|$ are often not reconstructed. This particularly effects $\tilde{\ell}$ with long lifetimes. Additionally, the p_{T} and $|d_0|$ and η cuts required to pass one of the triggers used in this analysis to pass the **LRT** filters further reduces the acceptance and efficiency.

The acceptance is highest for lifetimes around 0.1 ns, around 30% for $\tilde{\mu}$ and $\tilde{e}$ production and only 0.5% or $\tilde{\tau}$ production. The efficiency is higher, around 50% for lifetimes of order 0.1 ns for all flavors of $\tilde{\ell}$. The values for a range of possible mass and lifetime points can be see in Figs. 9.1, 9.2, 9.3, 9.4, and 9.5.

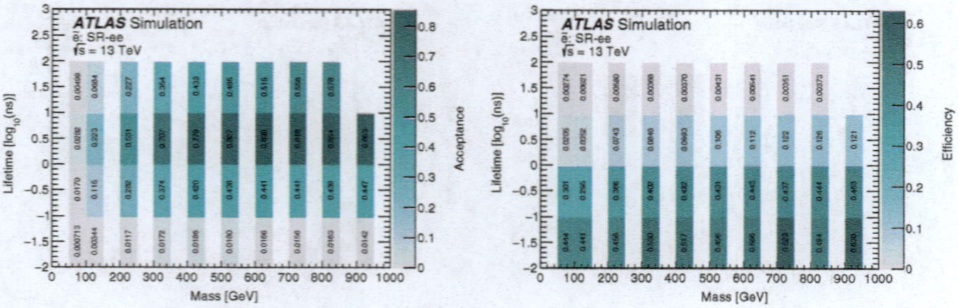

Fig. 9.2 Acceptance (left) and efficiency for $\tilde{e}$ decaying to electrons in SR-*ee*. The x-axis shows the possible masses of the $\tilde{e}$ and the y-axis its possible lifetime

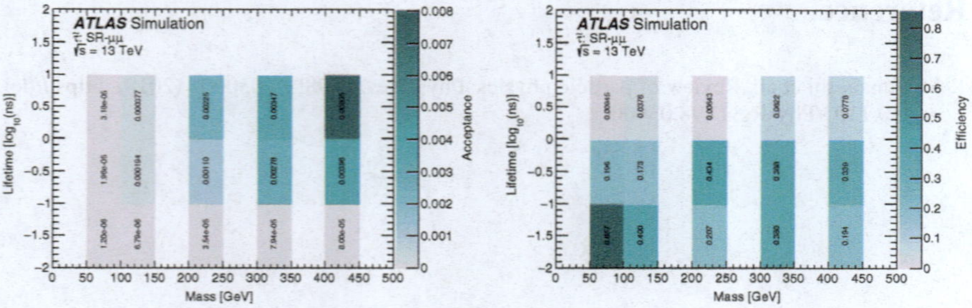

Fig. 9.3 Acceptance (left) and efficiency for $\tilde{\tau}$ decaying to electrons in SR-$\mu\mu$. The x-axis shows the possible masses of the $\tilde{\tau}$ and the y-axis its possible lifetime

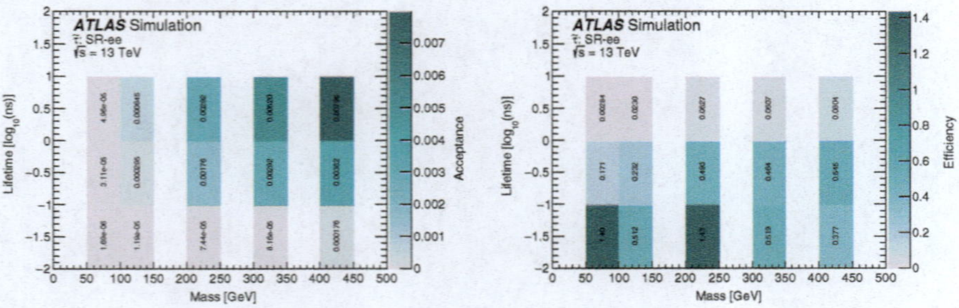

Fig. 9.4 Acceptance (left) and efficiency for $\tilde{\tau}$ decaying to electrons in SR-*ee*. The x-axis shows the possible masses of the $\tilde{\tau}$ and the y-axis its possible lifetime

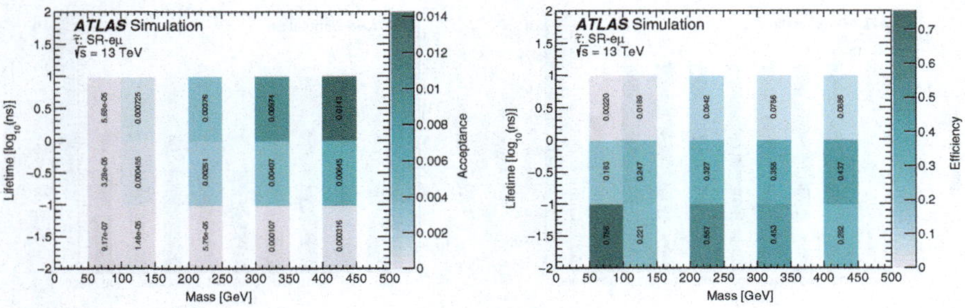

Fig. 9.5 Acceptance (left) and efficiency for $\tilde{\tau}$ decaying to electrons in SR-$e\mu$. The x-axis shows the possible masses of the $\tilde{\tau}$ and the y-axis its possible lifetime

Reference

1. M. Tanabashi et al., Review of particle physics. Phys. Rev. D **98**(3), 030001 (2018). https://doi.org/10.1103/PhysRevD.98.030001

Chapter 10
Backgrounds

In this analysis, backgrounds are estimated per signal region. In SR-*ee* and SR-$e\mu$, algorithmic fakes are the dominant source of background. In SR-$\mu\mu$, the background contribution from algorithmic fake muons and muons from heavy flavor decays is negligible ($\mathcal{O}(10^{-4})$), and the dominant background is from cosmic muons coincident with **LHC** collisions. While the signal lepton selection and event selection described in the previous chapter very efficiently remove these backgrounds, background estimates are calculated to estimate any residual contribution. All backgrounds are not well modeled in **MC**, so must be estimated from **CRs** in data, often resulting in statistical limitations.

10.1 Background to SR-*ee*

10.1.1 Fakes and Heavy Flavor Decays

The primary background to SR-*ee* is algorithmic fakes from the misassociation of a track with a real energy deposit in the **EM** calorimeter (such as from a photon); there is a secondary contribution from electrons from **HF** decays. **MC** samples of $t\bar{t}$ along with a photon dominated sample (described in Sect. 6.2) were used to study the relative contributions.

The $t\bar{t}$ provides a sample of **HF** decays (though not the dominant source of **HF** decays at the **LHC**), yet after all of the signal requirements, the remaining high $|d_0|$ electron was the result of a photon combined with an **ID** track. That this sample has many more b-hadron decays than photons, yet the photons contribute a larger background even in this narrow region of phase space, indicates that algorithmic fakes are the larger contributer of background events to SR-*ee*. Additionally, in truth-level studies of background **MC**, such as $BB \rightarrow \mu\mu$ (at truth level, the electron and muon kinematics are the same), $Z \rightarrow \tau\tau$ (where the τ decays include

L. Horyn, *A Search for Displaced Leptons in the ATLAS Detector*, Springer Theses,
https://doi.org/10.1007/978-3-030-91672-5_10

electrons), and $t\bar{t}$, all show exponentially falling distributions with no two lepton events that pass the p_T and $|d_0|$ cuts in the **SR**, as designed.

Further studies were performed in data, with one baseline electron passing the filter requirements and another lepton with no $|d_0|$ cut made. This second lepton must be *anti-isolated*, meaning it fails the isolation requirement made on signal leptons and has substantial activity surrounding it in the calorimeter and/or the **ID** and is likely to be inside of a jet. While the $\Delta p_T/p_T$ cut is designed to remove algorithmic fakes, it is a very effective remover of anti-isolated electrons as well. Clusters associated to electrons reconstructed inside of jets are likely to have additional energy incorrectly added to their clusters, increasing the cluster p_T and decreasing the $\Delta p_T/p_T$. Thus, it is not possible to disentangle heavy flavor electrons from fake electrons and the background contributions are estimated together with fakes as the dominant contribution.

10.1.2 Background Estimate

Fake electrons result from the failure of the reconstruction algorithm, and so the two fake electrons in an event should be uncorrelated. This assumption is used to estimate the background with an *ABCD method*. The procedure divides events into four regions based on the quality of the leading and subleading leptons in the event, shown in Fig. 10.1. The number of events in regions B, C, and D are combined to estimate the number of events in region A, the signal region:

$$N_A = \frac{N_B}{N_D} \times N_C \tag{10.1}$$

In the nominal estimate, region A is the signal region, with two electrons passing all signal cuts, region D has both electrons failing at least one signal cut, and regions C and D have only one electron passing all signal cuts while the other fails at least one (regions B, C, and D compose CR-*ee*-fake). For this estimate, a "failing" electron fails one of the $\Delta p_T/p_T$, χ^2_{ID}, or N_{miss} requirements, while a "passing" electron is a signal electron, passing all three cuts. The number of events in each region and the estimated number of events in SR-*ee* is shown in Table 10.1.

10.1.3 Validation and Systematic Uncertainties

The definitions of the ABCD regions can be varied to perform validations of the estimate and quantify systematic uncertainties. There are two ways to do this: first, the estimate can be done in slightly different ways, and the difference from the nominal estimate taken as an uncertainity; second, an estimate can be done in a validation region, where one or more signal cuts are inverted, and the estimated

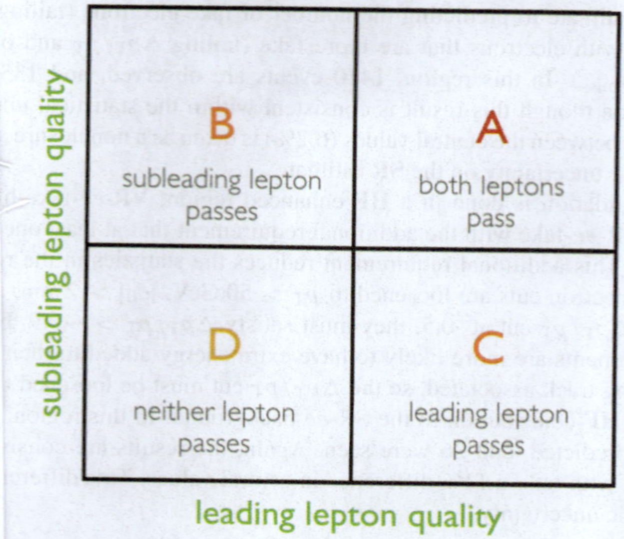

ɪe regions used for ABCD estimation

Results of the ABCD estimate for the nominal SR-*ee* estimate, the number of events
n are shown as well as the estimate for A. The uncertainties are statistical only and
ı statistics

	Nominal	Only $\Delta p_{\mathrm{T}}/p_{\mathrm{T}}$
	9068	1440
	77	28
	54	19
	0.459 ± 0.082	0.37 ± 0.11

events can be compared to the actual number of events in region A. In scenario, one looks for *closure*, that the estimate correctly estimates the ɜvents (within statistical uncertainties), giving evidence that the method ımption are sound. If this is not the case, the *nonclosure*, the extent to stimate disagrees with the correct number of events, is evaluated and an can be taken to cover this.

ɔns are performed by estimating the number of events in SR-*ee* in differ-For example, only the $\Delta p_{\mathrm{T}}/p_{\mathrm{T}}$, the most effective fake discriminator, the "passing"/"failing" variable. The results of this estimate are also 'able 10.1, the results are consistent within statistical uncertainties but l estimate is more precise.

ɔnd kind of validation is performed in two regions: one that enhances the ɔution, and a second that enhances the **HF** contribution. First, in a fake egion, VR-*ee*-fake, where the $\Delta p_{\mathrm{T}}/p_{\mathrm{T}}$ cut is inverted and the estimate ːd with χ^2_{ID} and N_{miss} together as the "passing"/"failing" variables. This changes the estimate to predicting the number of fake electrons (failing $\Delta p_{\mathrm{T}}/p_{\mathrm{T}}$) from regions with electrons that are more fake (failing $\Delta p_{\mathrm{T}}/p_{\mathrm{T}}$ and one or both of χ^2_{ID} and N_{miss}). In this region, 1440 events are observed, and 1356 ± 49 are predicted. Even though this result is consistent within the statistical uncertainties, the difference between the central values (6.2%) is taken as a nonclosure and applied as a systematic uncertainty on the **SR** estimate.

Next, a validation is done in a **HF** enhanced region, VR-*ee*-fake-hf, which is identical to CR-*ee*-fake with the additional requirement that at least one electron is anti-isolated. This additional requirement reduces the statistics in the region quite a bit, so the electron cuts are loosened to $p_{\mathrm{T}} > 50$ GeV, $|d_0| > 2$ mm, and instead of the usual $\Delta p_{\mathrm{T}}/p_{\mathrm{T}}$ cut at -0.5, they must satisfy $\Delta p_{\mathrm{T}}/p_{\mathrm{T}} > -0.9$. Electrons in dense environments are more likely to have extra energy added to their clusters or have the wrong track associated, so the $\Delta p_{\mathrm{T}}/p_{\mathrm{T}}$ cut must be loosened to probe the subdominant **HF** contribution to the SR-*ee* background. In this region, $23.5 \pm 1.$[illegible] events were predicted, and 26 were seen. Again, the results are consistent within uncertainties, with only a 11% difference in central values. This difference is taken as a systematic uncertainty.

10.1.4 Summary

This estimate is dominated by statistical uncertainties with additional, conservative systematic uncertainties taken, giving a final estimate of 0.46 ± 0.10 (0.082 stat. and 0.058 syst). This estimate has the smallest uncertainty of the three **SRs** due to the sufficient statistics in the estimate regions.

10.2 Background to SR-$e\mu$

10.2.1 Fake Background

The background to SR-$e\mu$ is very similar to the background in SR-*ee* and is estimated in a similar way. By tagging a lepton that fails either isolation or quality requirements and studying the properties of the other, probe lepton, the main contributing background (either fake or **HF**) is determined. Of the probe leptons, 100 pairs failed some signal requirements, but none only failed isolation, indicating that as in the case of SR-*ee*, algorithmic fakes are the dominant background to SR-$e\mu$.

Fake muons are rare compared to fake electrons due to the lack of extraneous activity in the **MS** compared to the calorimeter where photons add ambiguity to electron reconstruction. As a result, this estimate is extremely statistically limited.

10.2.2 Background Estimate

As in SR-ee, the two fake leptons in the event should be uncorrelated and so an ABCD method is used to estimate the background. A "failing" electron (as in SR-ee) is one that fails any one of the $\Delta p_T/p_T$, χ^2_{ID}, or N_{miss} requirements, a "failing" muon fails any one of the χ^2_{ID}, χ^2_{CB}, N_{miss}, N_{prec}, or N_ϕ requirements, and in both cases "passing" indicates a signal electron or muon.

However, when performing this estimate, the B region (electron passes, muon fails) has only 1 event, and the C region (muon passes, electron fails) has 0 events. This result cannot be used to calculate a background estimate, but it can be used to place an upper bound by setting the number of events in the C region to 1 event. The total number of events in SR-$e\mu$ must be less than 0.012 ± 0.017, where the uncertainty is statistical only. In order to increase the statistical power, different combinations of "passing" and "failing" leptons were required to only pass the baseline kinematic cuts of $p_T > 50$ GeV and $|d_0| > 2$ mm. Allowing both passing and failing leptons to only meet the baseline requirements allowed 1 event in the C region, enabling a background estimate of $0.007^{+0.018}_{-0.009}$. Statistical uncertainties are quoted using the Wilson interval for the C/D ratio summed in quadrature with the Poisson uncertainty on the number of events in the B region. This is taken as the nominal upper bound and the full result of this loosening can be seen in Table 10.2.

10.2.3 Validation and Systematic Uncertainties

As in the case of SR-ee, validations are performed enhancing the fake or **HF** contributions. VR-$e\mu$-fake again inverts the most powerful fake discriminators, $\Delta p_T/p_T$ for electrons and χ^2_{CB} for muons with the loosened p_T and $|d_0|$ cuts. In this region, 2 events are observed in the A region compared to $1.9^{+1.8}_{-1.0}$. While these

Table 10.2 Results of the ABCD method in the $e\mu$ channel in which the d_0 and p_T requirements are selectively loosened from 65 to 50 GeV, and 3 to 2 mm. The first column shows the results without any loosening, the second shows the results with the loosening applied only to the failing leptons, and the final column shows the results with the loosening applied in all regions. Uncertainties are statistical only. For the upper bound results, the value is obtained by setting the C region to 1 event, and the uncertainties are calculated using Poisson statistics. In the final case, the full calculation can be done

Estimate region	Signal p_T, $\|d_0\|$ cuts on all ℓ	Signal p_T, $\|d_0\|$ cuts passing ℓ	No signal p_T, $\|d_0\|$ cuts
D Observed	81	139	138
C Observed	0	0	1
B Observed	1	1	1
A Estimate	$<0.012 \pm 0.017$	$<0.007 \pm 0.010$	$0.007^{+0.018}_{-0.009}$

agree within the very large uncertainties, a 7.8% nonclosure systematic uncertainty is taken from the difference in central values.

Then, VR-$e\mu$-fake-hf is defined requiring at least one anti-isolated lepton with loosened p_{T} and $|d_0|$ cuts. To increase statistics, the $\Delta p_{\mathrm{T}}/p_{\mathrm{T}}$ cut is again loosened to -0.9 and the cuts on N_{prec} and N_{ϕ} are removed. Here, one event is observed in the A region, and $0.38^{+0.37}_{-0.32}$ events are predicted. To attempt this estimate another way with more statistical power, the estimate is performed again in this region, this time remaining agnostic to whether or not the lepton is isolated. This results in an estimate of $2.6^{+2.0}_{-1.4}$, while 5 events are observed. These numbers are again consistent within their substantial statistical uncertainties, but a conservative 92% nonclosure uncertainty is taken to account for the difference between the central values.

10.2.4 Summary

This region is extremely statistically limited such that a full background estimate is not possible, and so an upper limit is set. Less than $0.007^{+0.019}_{-0.011}$ ($^{+0.018}_{-0.009}$ syst. and 0.006 stat.) background events are expected in SR-$e\mu$.

10.3 Background to SR-$\mu\mu$

10.3.1 Cosmic Muon Identification

Muons from cosmic rays constantly pass through the earth and thus the **ATLAS** detector, particularly through the service shaft above the detector where there is no layer of earth above the detector. If a cosmic ray muon were coincident with a bunch crossing, the event could be triggered, reconstructed, and enter the dataset used for this analysis. The cosmic ray muon could pass through the entire detector at any distance from the **PV**, interacting with the **ID** and **MS**, and be reconstructed as two muons with high $|d_0|$, passing all quality variables (because the signature comes from a real muon) exactly mimicking the signature of SR-$\mu\mu$.

A very efficient cosmic tag was defined for this analysis to identify and remove muons from cosmic rays. This is done using a reoptimization of the strategy used by Ref. [1]. This method first defines a spatial cosmic identification, and then conservatively tags all muons which would be impossible to identify using this method due to gaps in detector coverage. The cosmic muon, μ_{cos}, passes through the entire detector and gets reconstructed as two muons, one on the top of the detector ($\phi > 0$) referred to as μ_{t} and the other on the bottom ($\phi < 0$), called μ_{b}. They follow the relationships

$$\Delta\phi = |\phi_{\mu_t} - \phi_{\mu_b}| = \pi \tag{10.2}$$

and

$$\Sigma\eta = |\eta_{\mu_t} + \eta_{\mu_b}| = 0 \tag{10.3}$$

The combination of these variables form a useful variable to describe events with 2 cosmic muons.

$$\Delta R_{\cos} = \sqrt{((\phi_{\mu_t} - \phi_{\mu_b}) - \pi)^2 + (\eta_{\mu_t} + \eta_{\mu_b})^2} \tag{10.4}$$

and it is useful to define

$$\Delta\phi_{cos} \equiv (\phi_0 - \phi_1) - \pi \tag{10.5}$$

and

The muon reconstruction algorithm described in Sect. 5.3 uses the momentum direction measured in the **MS** in its extrapolation from **MS** track to **ID** track and since μ_t is coming into the collision point instead of away from it, μ_t is often not fully reconstructed. In 90% of cases only μb is reconstructed and identified, while the detector signature for μ_t exists but the fully reconstructed muon does not. Thus it is advantageous to tag a cosmic muon as one which is back to back with activity in the **MS**, not as two muons back to back, illustrated in Fig. 10.2. A cosmic veto is defined based on the $\Delta\phi_{\cos}$ and $\Sigma\eta$ between a muon and a **MS** segment.

Since the **MS** is so far from the **PV**, the z_0 of individual MS segments is not measured and they are defined to point back to the origin. This creates a mismatch between the η of the segment and the η of the reconstructed muon

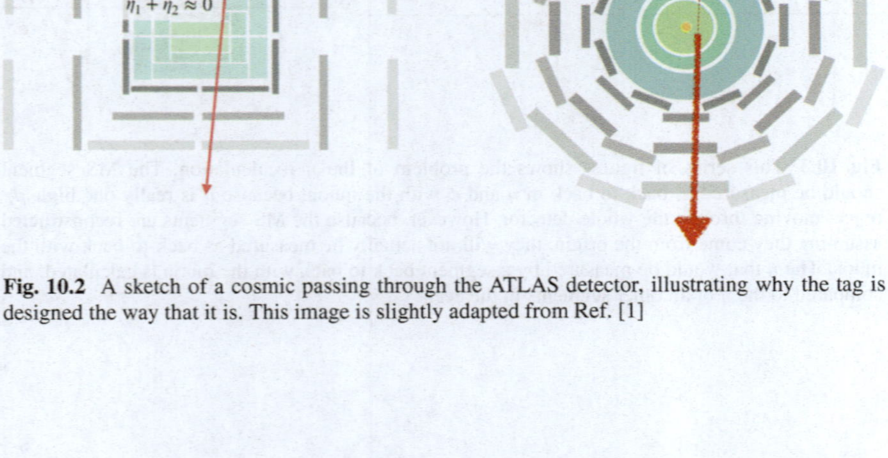

Fig. 10.2 A sketch of a cosmic passing through the ATLAS detector, illustrating why the tag is designed the way that it is. This image is slightly adapted from Ref. [1]

in the determination of the cosmic tag. This is geometrically corrected for by re-calculating the $\Delta\eta$ between the segment and the muon in the cosmic veto calculation. A schematic of this correction is shown in Fig. 10.3, its effect on the η and $\Sigma\eta$ measurements of muons in data is shown in Fig. 10.4, and its effect on the $\Delta\phi_{\text{cos}}$-$\Sigma\eta$ distribution of cosmic and signal muons shown in Fig. 10.5. This narrows the distribution of cosmic muons by an order of magnitude in $\Sigma\eta$. This distribution is isotropic for signal muons, so this definition allows for high cosmic muon rejection with minimal signal rejection.

The distribution the $\Delta\phi_{\text{cos}} - \Sigma\eta$ distribution is much narrower in $\Sigma\eta$ than in $\Delta\phi_{\text{cos}}$. This is because the **MS** measures η with an extremely high precision in the **MDTs** ($\mathcal{O}(10\ \mu\text{m})$), since it is the bending direction of the toroid and thus gives the momentum measurement, while the ϕ is measured by the **RPCs** with an order of magnitude less precision ($\mathcal{O}(10\ \text{mm})$). A high precision ϕ measurement of the combined muon comes from the **ID**, but the cosmic tag must contend with these resolution limitations to find a muon segment.

Additionally, there are gaps in the **MS** to allow for detector access, so a muon is conservatively tagged as a cosmic if it is back to back with this gap in detector coverage, as it could not be tagged as a cosmic using the geometric algorithm. A map of the material of the **MS** is used to veto cosmic muons using this *detector coverage veto*. Figure 10.6 shows the impact of each step on the muon distribution in VR-μ, which is dominated by muons from cosmic rays.

For this analysis, two different definitions of a "cosmic muon" are used. The *nominal tag* is used to veto events in the **SR**. If any baseline muon is tagged by the nominal cosmic tag, the entire event is vetoed. A second *narrow tag* is used during the validation of the estimate of background from cosmic muons. Both tags require muons to pass the detector coverage veto. Table 10.3 describes the values used for the various cosmic tags used and Fig. 10.7 shows the cuts used in the $\Delta\phi_{\text{cos}}$-$\Sigma\eta$ distribution.

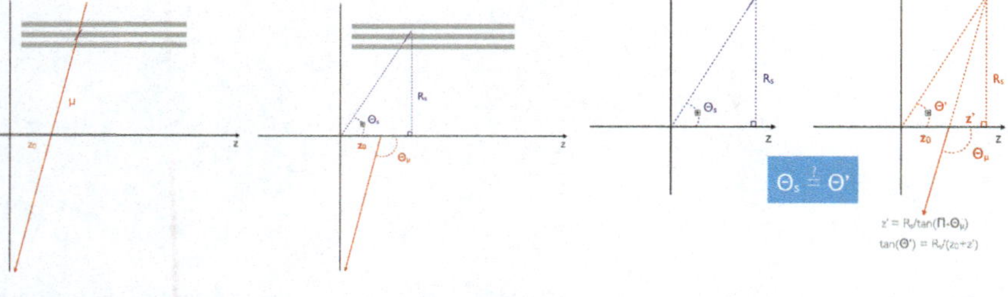

Fig. 10.3 This series of figures shows the problem of the η recalculation. The MS segment should be measured as back to back in η and ϕ with the muon, because it is really one high p_T object moving through the whole detector. However, because the MS segments are reconstructed assuming they come from the origin, they will not actually be measured as back-to-back with the muon. The η that would be measured by a segment back to back with the muon is calculated, and compared to the η of all other segments in the event

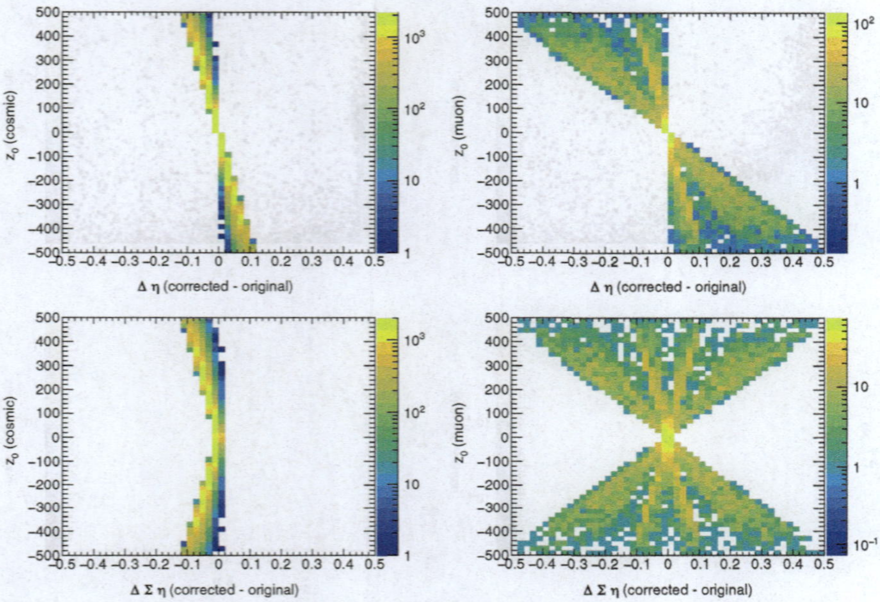

Fig. 10.4 Change to the η measurement (top) and $\Sigma\eta$ measurement (bottom) due to the z_0 correction. These plots compare muons tagged with due to the presences of a segment in the $\Delta\phi_{\cos} - \Sigma\eta$ window (left) to those untagged or tagged with the detector coverage veto (right). The behavior shows isotropic effect for the muons which are not tagged via the algorithm that uses this correction, while the $\Sigma\eta$ distribution of cosmic muon always decreases due to the correction

Table 10.3 Cuts applied in full and narrow cosmic tags. The narrow is contained in the full tag. An intermediate tag is defined as the region between the full and narrow tags, that is, tagged by the full but not the narrow tag

Tag	$\Delta\phi(\mu, \text{seg})$	$\Sigma\eta(\mu, \text{seg})$	Pass coverage acceptance	Rejection efficiency
Full	<0.25	<0.018	True	99.5%
Narrow	<0.02	<0.013	True	59%

The cosmic tagging efficiency is determined by fitting distributions in VR-μ. A template of the t_0^{avg} of cosmic tagged muons and prompt (collision) muons is taken from data shown on the left of Fig. 10.8. Any correlation between t_0^{avg} and d_0 can not be observed within detector resolutions, so the prompt sample of muons is taken to represent the signal muons that would result from collisions.

These are used to determine the fraction of cosmic muons in VR-μ before and after the cosmic tagged muons are removed from VR-μ (left of Fig. 10.8). From the fraction of cosmic muons estimated from the template fitting, the fraction of muons removed by the cosmic tag can be measured. The nominal tag removes 99.5% of cosmic muons and the narrow tag removes 59% of cosmic muons. The nominal

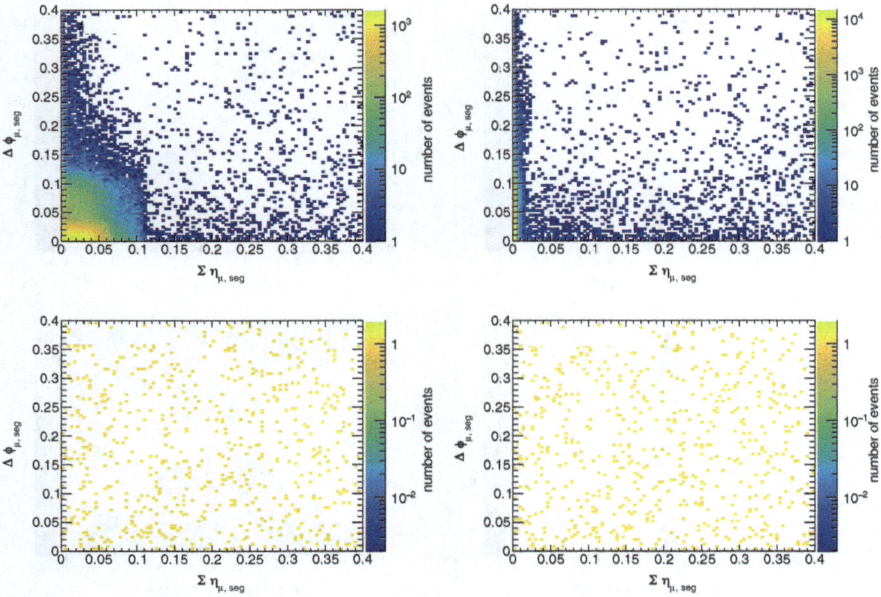

Fig. 10.5 The $\Sigma\eta - \Delta\phi$ distribution is shown before (left) and after (right) the η recalculation. The top row shows the distributions in VR-μ and bottom row 300 GeV $\tilde{\ell}$ signal samples in our full range of lifetimes. Using this cosmic tag, a very tight cut can be made on cosmic muons without losing signal efficiency

tag removes 8% of collision muons, primarily from the detector coverage veto. Figure 10.9 compares the kinematic distributions and cosmic-tagging efficiency between VR-μ and signal **MC**.

10.3.2 Properties of Events with Cosmic Tagged Muons

Because all events with a cosmic tagged muon are vetoed, a **CR** with at least one cosmic tagged muon, CR-M_{full}, is defined to study cosmic events. There are about 224,000 events with at least one cosmic-tagged muon in the full dataset. 90% of events in CR-M_{full} have only one muon, and that muon is cosmic tagged. 90% of those events find the cosmic muon on the bottom of the detector, μ_{b}. The other 10% of events have two reconstructed muons and they are both cosmic tagged. One event was seen that had four cosmic tagged muons, two on either side of the detector, indicating muons from a cosmic shower. In CR-M_{full} events, some jets are seen from pileup collisions, but no additional leptons are observed in events with cosmic muons. Interesting **LHC** collisions are rare and so are cosmic muons, so the odds of

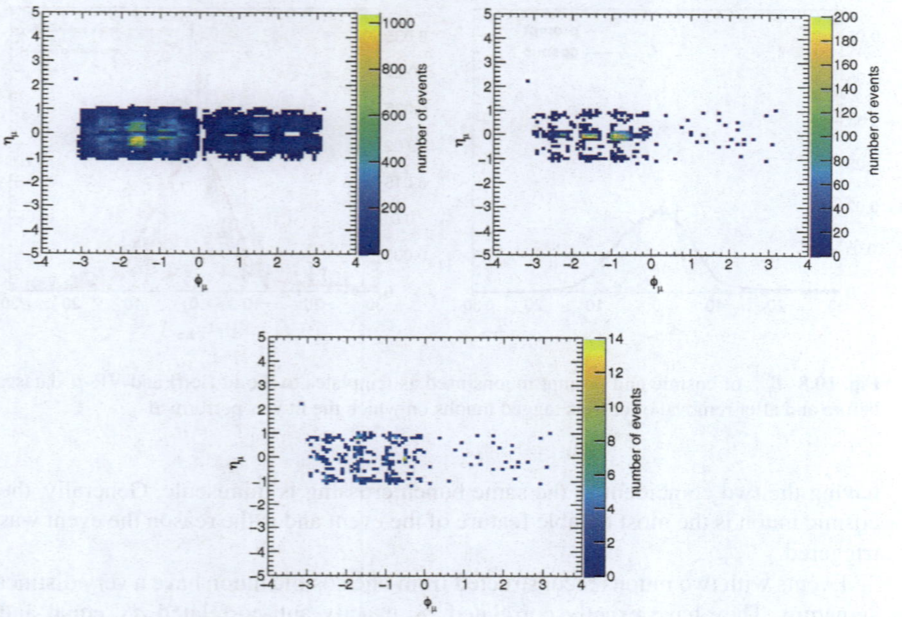

Fig. 10.6 The $\eta - \phi$ distribution of muons in VR-μ are shown after baseline cuts (top left), the cosmic veto (top right), then the detector coverage veto (bottom)

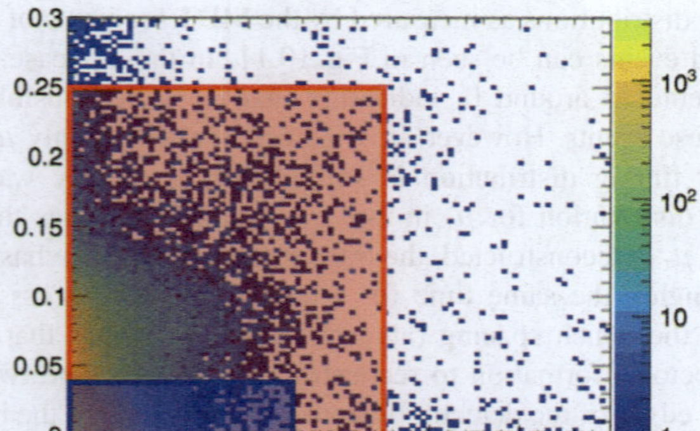

Fig. 10.7 The $\Sigma\eta_{\mu}$,seg $- \Delta\phi_{\mu}$,seg distribution of signal quality muons in VR-μ. The red line and shadow shows the bounds of the full cosmic tag, while the blue line and shadow shows the boundary of the narrow cosmic tag. Everything inside the respective boxes is tagged as a cosmic. The intermediate tag defines the region included in the full tag, but not included in the narrow tag

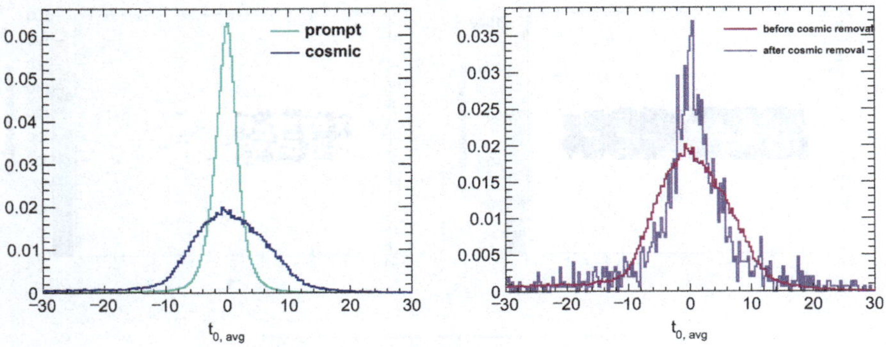

Fig. 10.8 t_0^{avg} of cosmic and prompt muons used as templates to the fit (left) and VR-μ dataset before and after removal of cosmic tagged muons on which the fit was performed

having the two coincident in the same bunch crossing is minuscule. Generally, the cosmic muon is the most notable feature of the event and is the reason the event was triggered.

Events with two muons reconstructed from one cosmic muon have a very distinct signature. They have exactly correlated z_0, exactly anti-correlated d_0, equal and opposite ϕ measurement, and their η measurements sum to 0. This can be seen in Fig. 10.10. No events were found with evidence of a cosmic muon's trajectory changing due to radiation in the detector, so this effect is assumed to be negligible within the detector resolution.

The timing distributions as measured by the MDT segments of muons in 1 and 2 cosmic tagged events can be seen in Fig. 10.11. In the 1 μ case, μ_{b} has a timing distribution centered around 0, indicating that μ_{b} is responsible for the trigger decision in these events. However, in 1-μ events in which only μ_{t} passes baseline selections, its timing distribution is shifted negative (early w.r.t the collision), mirroring the distribution for μ_{t} in 2-μ events. This indicates that, even in cases in which only μ_{t} is reconstructed, the trigger decision is made based on μ_{b}.

It takes roughly the same time for a cosmic muon to cross the width of the detector as is the bunch spacing (about 25 ns). This means that in order to have sufficient detector information to reconstruct two muons, the two **MS** signatures must be at the edges of the detector readout window and near the bounds of the t_0^{avg} cut. This makes one or both muons likely to have detector information associated to the wrong event. Due to the early timing of μ_{t}, this is more likely for μ_{t}, but occurs for both μ_{t} and μ_{b}.

Each of the combined muons will pass signal selections, because high quality information exists from the **ID**, but if enough **MS** information is missing, the **MS** segments will not be found back to back with the opposite combined muon, causing one or both of the muons to evade the cosmic tag. Ultimately, it is this mismeasurement that leads to the background in SR-$\mu\mu$.

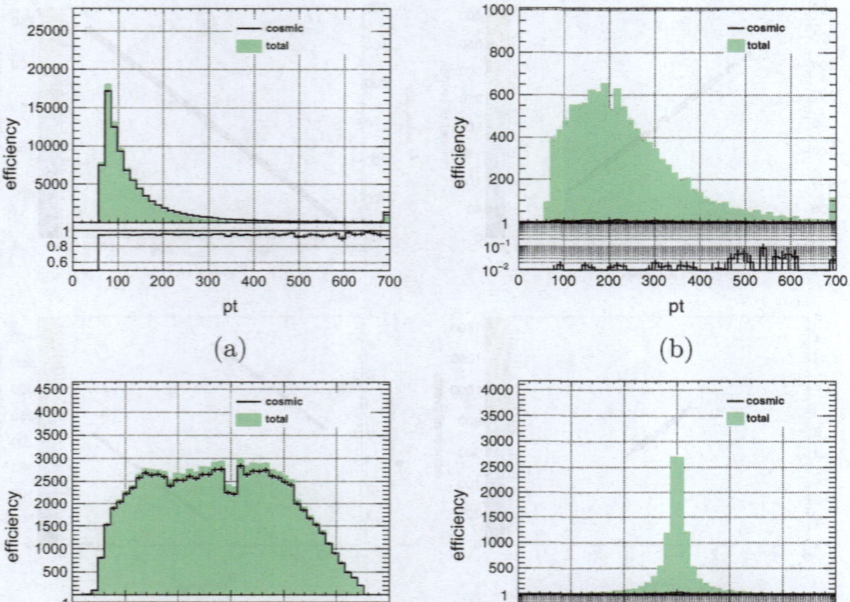

Fig. 10.9 Cosmic tagging efficiency with respect to kinematic variables in VR-μ data and signal **MC**. (**a**) VR-μ data with respect to p_{T}. (**b**) Signal **MC** with respect to p_{T}. (**c**) VR-μ data with respect to d_0. (**d**) Signal **MC** with respect to d_0. (**e**) VR-μ data with respect to z_0. (**f**) Signal **MC** with respect to z_0. (**g**) VR-μ data with respect to ϕ. (**h**) Signal **MC** with respect to ϕ. (**i**) VR-μ data with respect to η. (**j**) Signal **MC** with respect to η

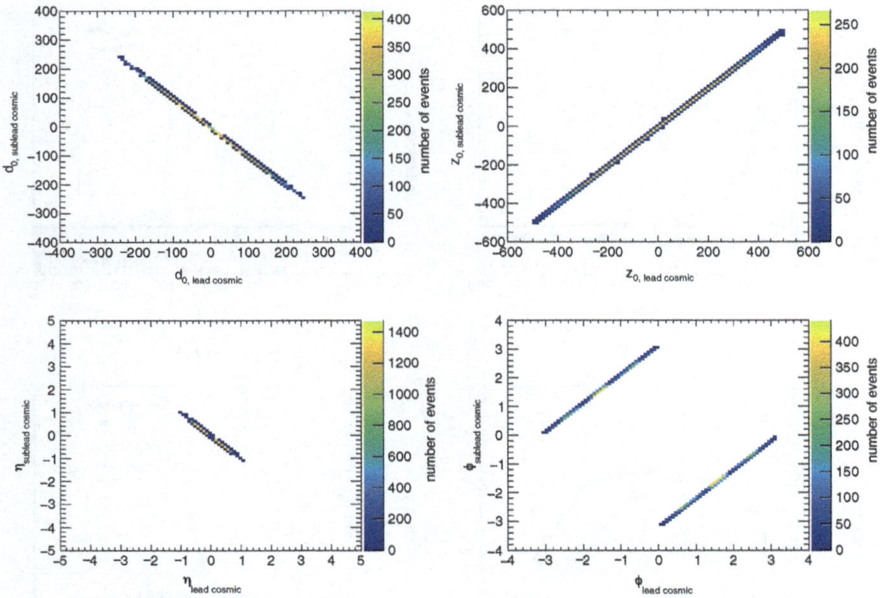

Fig. 10.10 Relationship between two cosmic tagged muons in an event. Their d_0 (top left), z_0 (top right), η (bottom left), and ϕ (bottom right) values all indicate that the two muons originate from the same cosmic

10.3.3 Background Estimate

If a **MS** segment is reconstructed without direct ϕ measurements from **RPC** or **TGC** hits, its ϕ measurement is taken as the center of the **MDT**, which has resolution of $\Delta\phi = 0.2$ (the final nominal tag value of $\Delta\phi_{\text{cos}} = 0.25$ was expanded to include these cases). Events with this measurement scheme were seen in a preliminary definition of the cosmic tag, which used the same $\Sigma\eta$ cut, but the $\Delta\phi_{\text{cos}}$ cut was reduced to $\Delta\phi_{\text{cos}} = 0.18$. With this preliminary version of the cosmic tag, about 40 events were observed with one cosmic tagged muon, and another untagged muon. In these events, μ_{t} was missing direct **MS** ϕ measurements, so its segments were not back-to-back with μ_{b} and μ_{b} was not tagged as a cosmic. μ_{b} however, was well measured, so μ_{t} was cosmic tagged. An event can enter the signal region if both μ_{b} and μ_{t} are sufficiently mis-measured that neither can be tagged. This is sketched in Fig. 10.12. A muon's cosmic tag is dependent on the opposite muon's quality, so the muon's cosmic tag status and its quality are assumed to be uncorrelated in order to make an estimate of the background.

This estimate and validation makes use of the two cosmic tags described in Table 10.3, as well as an *intermediate tag* which defines a muon that is not tagged by the narrow tag but is tagged by the nominal tag. Because a cosmic muon is defined using detector information on the opposite side of the detector, muons tagged using

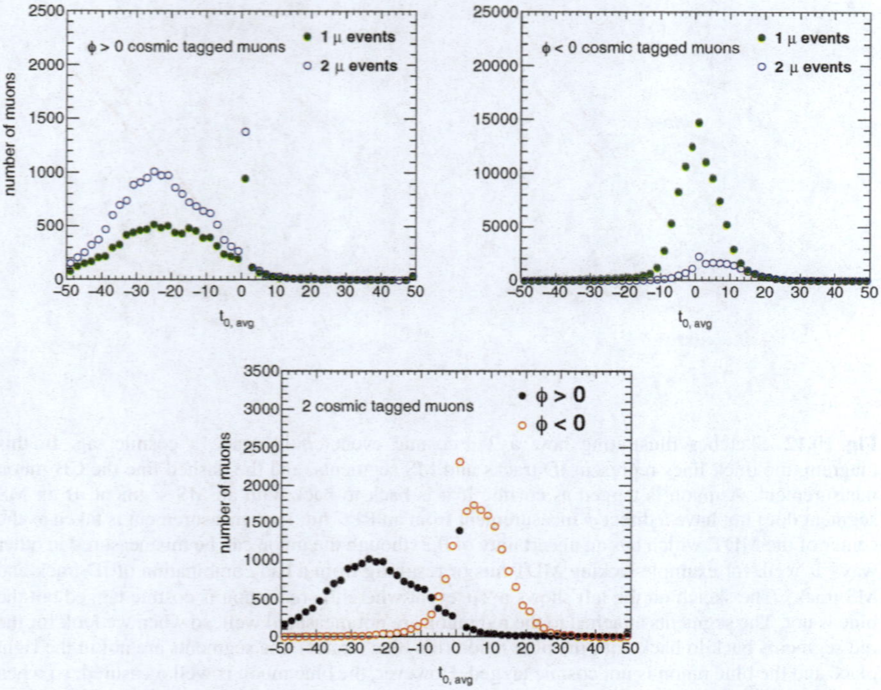

Fig. 10.11 Comparison of timing distributions of positive ϕ (top left) and negative ϕ (top right) muons in 1 and 2 cosmic events, as well as the timing distribution of top and bottom muons in 2 cosmic events (bottom). Here, "2 cosmic tagged muons" implies 2 muons reconstructed from 1 cosmic muon. t_0^{avg} is calculated by taking the average of the t_0 measured by all segments associated to the muon. Note that the peak at $t_0^{avg} = 0$ indicates a failure of the fit used to measure t_0^{avg}, these are handled separately and individual segments with $t_0^{avg} = 0$ do not enter the t_0^{avg} calculation

the narrow tag must have a higher quality measurement on the opposite side of the detector in order to pass the tighter cuts than those tagged with only the nominal tag. Cosmic tagged muons with $\Delta\phi_{\cos} - \Sigma\eta$ values closer to the edge of the nominal cosmic tag window are more similar to those that would enter SR-$\mu\mu$ as background (which are outside of the cosmic tag window).

To estimate the number of events entering SR-$\mu\mu$ from poorly measured cosmic muon events, a scaling factor from good quality to bad quality muons, R_{good}, is defined. Then, it is used to scale events which have one good quality and one bad quality muon, CR-$\mu\mu$-topbad, into SR-$\mu\mu$ to estimate the background contribution. This is in Fig. 10.13. "Good" quality defines a muon that passes signal cuts on N_{prec}, N_{ϕ}, and χ^2_{CB}, while a "bad" quality muon fails at least one of these cuts. For the nominal version of the estimate, it is assumed that there is one $\phi > 0$ muon (μ_{t}) and one $\phi < 0$ muon (μ_{b}), and μ_{t} is the muon is scaled from "bad" to "good" quality.

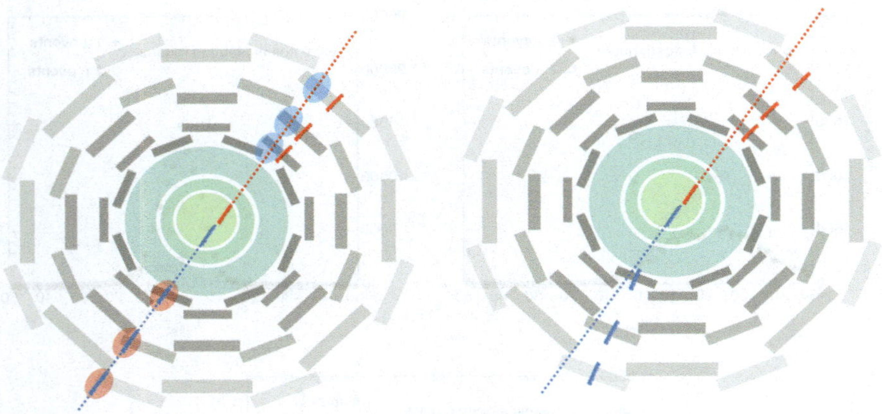

Fig. 10.12 Sketches illustrating how a 2 μ cosmic event could evade a cosmic tag. In this diagram, the thick lines represent ID tracks and MS segments, and the dashed line the CB muon measurement. A muon is tagged as cosmic if it is back to back with an MS segment. If an MS segment does not have a direct ϕ measurement from an RPC hit, its ϕ measurement is taken as the center of the MDT, which has an uncertainty of 0.2 (though the muon can be mismeasured in other ways as well, for example lacking MDT hits or resulting from a bad combination of ID track and MS track). The sketch on the left shows a 2 μ event where the red muon is cosmic tagged but the blue is not. The segments attached to the red muon are not measured well, so when we look for the red segments back to back with the blue muon (the blue circles), the segments are not in the right place and the blue muon is not cosmic tagged. However, the blue muon is well measured, so when we look back to back with the red muon, we find the segments of the blue muon. Thus, the better quality muon is not tagged, while the poorer quality muon is. The right shows a scenario where both muons have this mismeasurement, and so neither is cosmic tagged. This is what contributes to the background in SR-$\mu\mu$

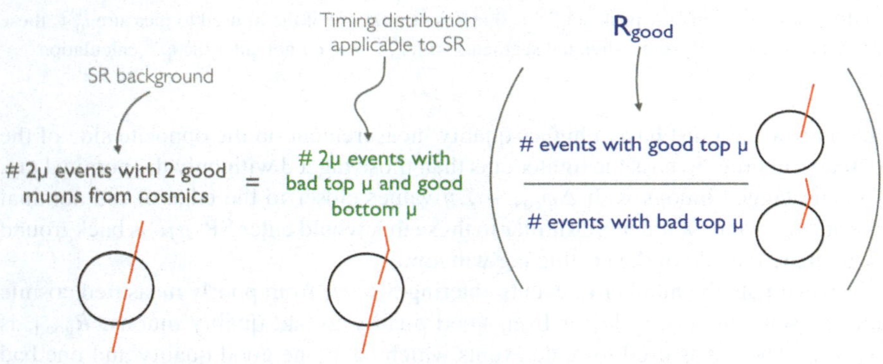

Fig. 10.13 A visual representation of the background estimation strategy

All regions have 2 reconstructed muons, and the cosmic tag and quality are varied to define the various regions used to make the estimate (Table 10.4).

Table 10.4 A description of the regions used for the cosmic estimate and validation. Each column describes the way in which the muon deviates from a signal muon, meaning a muon is signal in all respects except for the parameter(s) listed in the table

Region	$\phi < 0$ muon	$\phi > 0$ muon
1	Signal	Signal
2	Signal	Fails at least one MS quality cut
3	Signal	Cosmic tagged
4	Signal	Cosmic tagged and fails at least one quality cut
5	Narrow tagged	Narrow cosmic tagged
6	Narrow tagged	Narrow cosmic tagged and fails at least one quality cut
7	Narrow tagged	Full, but not narrow, tagged
8	Narrow tagged	Full, but not narrow, tagged and fails at least one quality cut

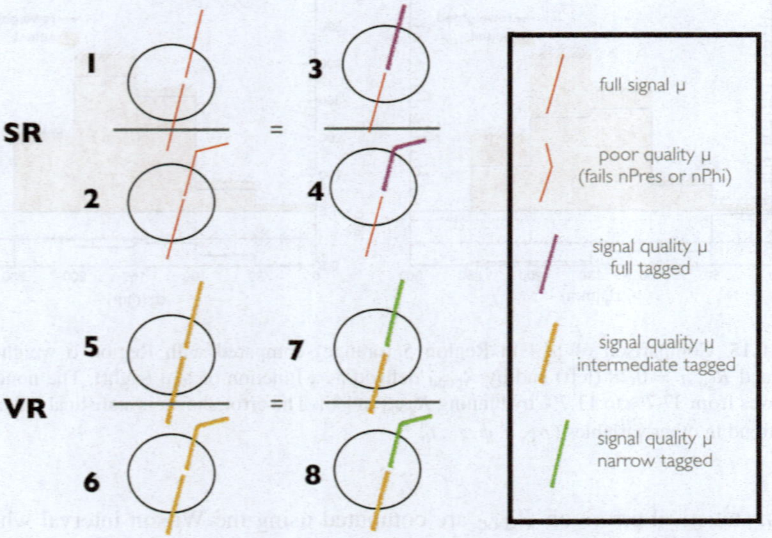

Fig. 10.14 A visual representation of the CRs and VRs used to estimate and validate the background contribution to SR-$\mu\mu$ due to cosmic muons. Regions 3 and 4 are used to define R_{good}. These regions were chosen to maintain orthogonality, while minimizing signal contamination and maximizing statistical power

R_{good} for the SR-$\mu\mu$ estimate is defined using muons tagged with the full cosmic tag. A validation estimate is also defined, using the narrow and intermediate cosmic tags. The signal and validation regions used in this estimate are sketched in Fig. 10.14 and the numbers of events in each region listed in Table 10.5. Aside from Region 2 (CR-$\mu\mu$-topbad), all other regions in the SR and VR estimates are subsets of CR-M_{full}. These regions were chosen to maintain orthogonality, while minimizing signal contamination and maximizing statistical power. In all

Table 10.5 Numbers of events in each region used for the cosmic estimate. Regions 2–4 are used to estimate SR-$\mu\mu$ (Region 1) and Region 6–8 estimate the number of events in Region 5, show in Fig. 10.14

Region	1	2	3	4	5	6	7	8
Event yield	–	2	1	18	1088	1000	1947	2465

Table 10.6 Results of the background estimation strategy in the two validation regions and the signal region

Region	Central value	Up error	Down error	Actual value	% Diff from actual
VR	789.9	34.85	34.4	1088	37.7%
SR	0.11	0.20	0.10	–	–

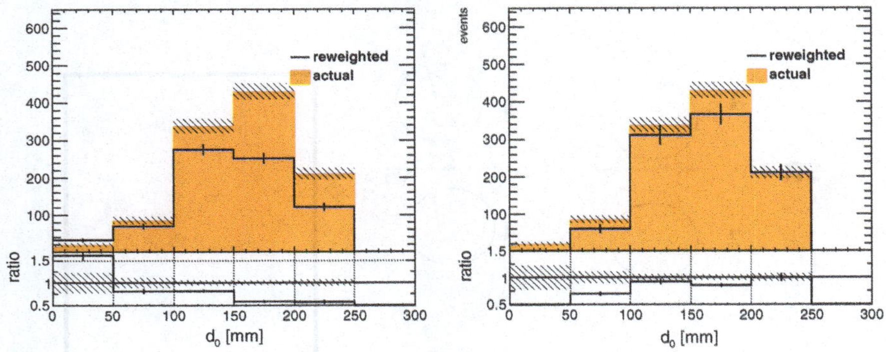

Fig. 10.15 Comparison of $|d_0|$ in Region 5 (orange) compared with Region 6 weighted by unbinned R_{good} $= 0.78$ (left) and by R_{good} defined as a function of $|d_0|$ (right). The nonclosure improves from 37.7% to 11.7% by binning R_{good} in $|d_0|$. The error shown is statistical only. There is no trend in other variables (p_{T}, η, ϕ, z_0, t_0^{avg})

cases, statistical errors on R_{good} are computed using the Wilson interval which is recommended for values close to zero with small statistics [2].

Results of the estimate and validation are shown in Table 10.6. There is a statistically significant discrepancy between the estimated and actual number of events in the VR. This nonclosure can be explained by the difference in $|d_0|$ distributions in Region 5 and Region 6, shown in Fig. 10.15. The excess of low $|d_0|$, high χ^2_{CB} muons in the weighted Region 6 compared to Region 5 implies that more muons have been created due to a poor combination of prompt track with MS track. The estimate was performed in the VR with R_{good} defined as a function of $|d_0|$ and the nonclosure was reduced from 37.7% to 11.7%. It is not possible to perform this estimate as a function of $|d_0|$ in the SR due to low statistics in Region 2 and Region 3. The estimate is performed without the $|d_0|$ binning and the nonclosure from the unbinned estimate in the VR is taken as an uncertainty (37%). In the SR, seen in Fig. 10.17, there are no overlapping $|d_0|$ bins in CR-$\mu\mu$-topbad and R_{good}, however, adjacent bins are filled, so the extrapolation over $|d_0|$ is not large and uncertainity

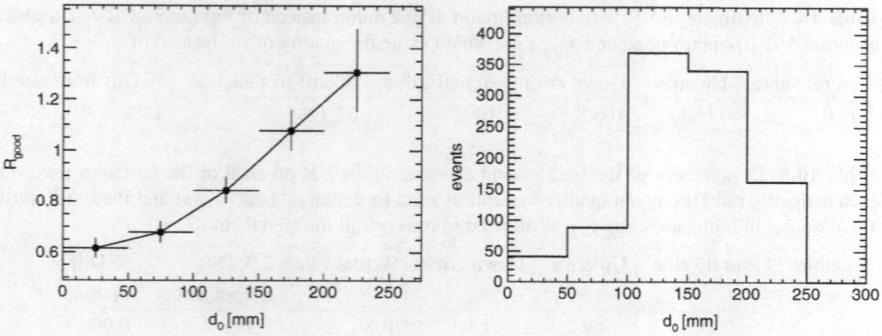

Fig. 10.16 $|d_0|$ distribution of R_{good} (left) and μ_t in Region 6 to perform the VR estimate. To calculate the background estimate, the distributions are multiplied bin by bin and summed. The error shown is statistical only. This method is ultimately not used, in favor of an unbinned version, due to statistical limitations in the regions used for the SR estimate

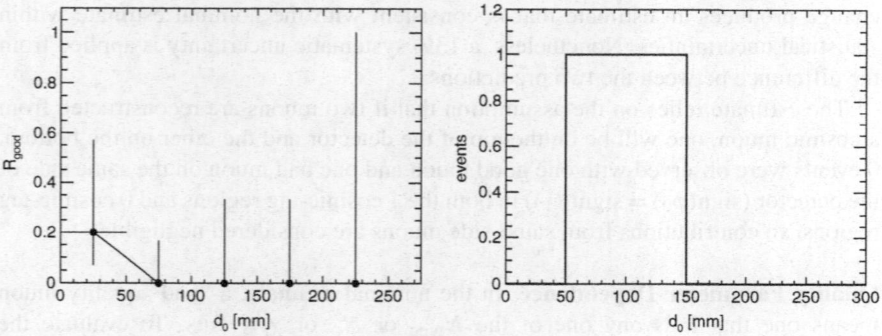

Fig. 10.17 $|d_0|$ distribution of R_{good} (left) and μ_t in Region 2 to perform the SR estimate. To calculate the background estimate, the distributions are multiplied bin by bin and summed. The error shown is statistical only. It can be seen here that there are no overlapping bins between the two plots, and so this method is ultimately not used, in favor of an unbinned version

from this extrapolation is contained in the nonclosure systematic. R_{good} and $|d_0|$ distributions for each estimate is shown in Figs. 10.16 and 10.17 for the SR and VR, respectively.

10.3.4 Systematic Uncertainties

Muon Orientation The estimate is performed using the quality of μ_t since muons in the upper hemisphere are expected to be more temporally marginal and mismeasured. However, the strategy can be performed with the quality of μ_b. Because bad quality μ_b are more rare bad poor quality μ_t, it is not possible to

Table 10.7 Estimate in VR with bottom muon as test muon instead of top muon. CR-$\mu\mu$-topbad becomes VR-$\mu\mu$-bottombad and R_{good} is defined using the quality of μ_b instead of μ_t

Central value	Up error	Down error	Actual value	% Diff from actual	% Diff from nominal
689.0	115.1	108.8	1088	62.3%	13%

Table 10.8 Dependence of the background estimate in the VR on each of the variables used. In each estimate, only the given quality variable is used to define a "bad" muon and the other must always pass. In both cases, the χ^2_{CB} is allowed to pass or fail the signal cut

Variable	Central value	Up error	Down error	Actual value	% Diff from actual	% Diff nominal
N_ϕ	783.6	69.2	68.1	1088	38.8%	0.9%
N_{prec}	920.6	51.8	50.9	1088	18.2%	16.5%

compare this strategy in the SR estimate, which is already statistically limited. However, the comparison can be made in the VR. As shown in Table 10.7, this change produces an estimate that is consistent with the nominal estimate within statistical uncertainties. Nonetheless, a 13% systematic uncertainty is applied from the difference between the two predictions.

The estimate relies on the assumption that if two muons are reconstructed from a cosmic muon, one will be on the top of the detector and the other on the bottom. 0 events were observed with one good muon and one bad muon on the same side of the detector ($\text{sign}(\phi_0) = \text{sign}(\phi_1)$) in both the 1 cosmic-tag regions and 0 cosmic-tag regions, so contributions from same-side muons are considered negligible.

Quality Parameter Dependence In the nominal estimate, a "bad" quality muon means one that fails any one of the N_{prec} or N_ϕ or χ^2_{CB} cuts. To evaluate the dependence one each variable, the estimate can be performed using only one of these to define a "bad" quality muon, requiring the others to pass the signal requirements. However, χ^2_{CB} is dependent on the MS track hit requirements and has a small contribution to the estimate regions on its own. Only inverting only χ^2_{CB} leaves 11 events in each Region 6 and Region 8 and 0 in Region 2 and Region 4. Thus, we account for the χ^2_{CB} contribution by remaining agnostic to it in performing the estimate with the other two quality variables. Allowing for the failure of the χ^2_{CB} cut increases the N_{prec}-only and N_ϕ-only estimates by 2% and 1%, respectively. Tables 10.8 and 10.9 show the results of this procedure in the VR and SR, respectively. We take the largest difference from the nominal estimate in the VR (16.5%) to account for the quality variable dependence.

Table 10.9 Dependence of the background estimate in the SR on each of the variables used. In each estimate, only the given quality variable is used to define a "bad" muon and the other must always pass. In both cases, the χ^2_{CB} is allowed to pass or fail the signal cut

Variable	Central value	Up error	Down error	% Diff from nominal
N_ϕ	0	–	–	100%
N_{prec}	0.33	0.75	0.40	200%

10.3.5 Summary

This estimate is dominated by statistical uncertainties in the SR estimation regions. Ultimately, $0.11^{+0.20}_{-0.11}$ ($^{+0.198}_{-0.104}$ stat. and 0.047 syst) events are expected in SR-$\mu\mu$ due to cosmic muons.

10.4 Negligible Backgrounds

Since this analysis is built on the assumption that there should be 0 background events in all signal regions, it is extremely important to ensure that all possible backgrounds are accounted for. Three different backgrounds were studied and determined to be negligible, meaning that they contribute $\mathcal{O}(10^{-4})$ or less events to the relevant **SRs**.

10.4.1 Material Interactions

Material interactions define events with dilepton pairs that come from interaction with the material of the **ATLAS** detector. In other searches for long lived particles in **ATLAS** where displaced vertices (**DVs**) are required, **DVs** are reconstructed by extrapolating their tracks to their intersection, and it can be determined if vertices originate from material. In this analysis, there is no vertex so such a veto cannot be used. In order to study material interactions, events with leptons associated to displaced vertices from material were studied. It was found that in the cases of both electrons and muons, events with an **DV**, where one track is associated to a lepton, the ΔR between the lepton and the other track in the **DV** was less than 0.2. After making the requirement that $\Delta R_{\ell,\mathrm{track}} < 0.2$, no **DVs** with signal leptons were found. In the case of the background to one of the **SRs**, the second track in the **DV** would be associated to a lepton. After this cut is made, this background is negligible to all 3 **SR**.

10.4.2 Fake Muons

Fake muons were estimated using an ABCD method similar to SR-ee and SR-$e\mu$. Muons were selected using the baseline requirements, plus asking that be isolated, and not cosmic tagged. All remaining cuts used for defining signal muons were used to define "passing" muons. Using this strategy, only 6 events were seen in the D region, and 0 in either B or C. In the SR-$e\mu$ fake estimate, the ratio of passing to failing muons is less than 1%. Given this, the probability for two fake muons to pass our selections should be less than 0.01%, making this background negligible (<0.0006 events) and we consider it to be negligible relative to cosmic muons.

10.4.3 Heavy Flavor Muons

Heavy flavor electrons and muons cannot be disentangled from algorithmic fakes in SR-ee and SR-$e\mu$ and are estimated in conjunction with fakes. Fakes are negligible in SR-$\mu\mu$, so a **HF** estimate must be performed separately. Very few events in $t\bar{t}$ **MC** have single muons from decays of b-hadrons pass the signal p_T and $|d_0|$ cuts, so this background is expected to be small when two muons are required. An ABCD method cannot be used for this estimate because it cannot be assumed that the probability to find an isolated muon is uncorrelated between 2 muons from two **HF** decays.

First, data was checked for anti-isolated muon events in CR-$\mu\mu$-hf, a region with two signal muons and at least one of the muons must be anti-isolated. No events were observed. Loosening the kinematic cuts to $p_T > 50$ GeV and $|d_0| > 2$ mm also found zero events. Finally, since the muon trigger and thus muon `DRAW_RPVLL` filter only requires one muon, it was possible to study events with one baseline muon and one muon with $|d_0| > 0.5$ mm. One event was observed in this region. Then, extrapolation factors to the **SR** kinematic cuts were defined from the same dataset. The extrapolation factor from the extremely loosened region into SR-$\mu\mu$ for both the $|d_0|$ and p_T of both muons was calculated to be 0.00013 ± 0.00013. This means that 1.3×10^{-4} are expected in SR-$\mu\mu$ from muons from **HF**, a negligible contribution.

10.5 Summary

In all, <1 events are expected in each **SR**. Table 10.10 summarizes the background estimates and Table 10.11 summarizes the systematic uncertainties.

Table 10.10 Summary table of the background estimate and uncertainty in each **SR**

Region	SR-ee	SR-$\mu\mu$	SR-$e\mu$
Total background	0.46 ± 0.10	$0.11^{+0.20}_{-0.11}$	$0.007^{+0.019}_{-0.011}$
Fakes + heavy flavor	0.46 ± 0.10	$<10^{-4}$	$0.007^{+0.019}_{-0.011}$
Cosmics	–	$0.11^{+0.20}_{-0.11}$	–

Table 10.11 Table describing statistical and systematic uncertainties for all background estimates as a percent of total yield. The total uncertainty is the sum of the individual components in quadrature

Background	Uncertainty	Value [%]
SR-ee		
Fake and heavy flavor	Statistical	18
	HF nonclosure	11
	Fake nonclosure	6
	Total	22
SR-$e\mu$		
Fake and heavy flavor	Statistical	+260/−130
	HF nonclosure	92
	Fake nonclosure	8
	Total	+270/−160
SR-$\mu\mu$		
Cosmic muons	Statistical	+180/−95
	VR nonclosure	38
	Estimate variable	16.5
	Muon orientation	13
	Total	+185/−104

References

1. ATLAS Collaboration, Search for long-lived, massive particles in events with a displaced vertex and a muon with large impact parameter in *pp* collisions at $\sqrt{s} = 13$ TeV with the ATLAS detector. Phys. Rev. D **102**(3), 032006 (2020). https://doi.org/10.1103/PhysRevD.102.032006. arXiv: 2003.11956 [hep-ex]
2. A. DasGupta, L.D. Brown, T.T. Cai, Interval estimation for a binomial proportion. Stat. Sci. **16**(2), 101–133 (2001)

Chapter 11
Signal Systematics

In order to use this analysis to make a statement about a potential **BSM** model, the extent to which the signal **MC** correctly simulates the real physical environment must be evaluated. Differences between data and **MC** are studied in order to avoid underestimating or overestimating the expected number of **BSM** events that could have been seen in the data.

Where possible, the **MC** is corrected to better represent the data using *scale factors*, and in other cases systematic uncertainties are applied to the final result. Uncertainties are also evaluated on the scale factors. A list of all of the systematic uncertainties applied in the interpretation are listed in Table 11.1. The value listed in the table describes how much varying the efficiency of a given parameter changes the final signal yield.

The dominant source of systematic uncertainty on the signal **MC** in this analysis comes from the efficiency for selecting displaced leptons, for which there is no data to compare to **MC**. As a result, this is evaluated in several steps: first the trigger, reconstruction, and selection efficiencies are compared for prompt leptons in the same physics process in data and **MC**; then the tracking efficiency is compared between signal muons and cosmic muons, which compares different physical phenomena that result in p_{T}, high $|d_0|$ tracks; finally, the lepton reconstruction efficiency is studied with respect to displacement is studied in **MC** only and a conservative additional uncertainty to account for any missed effects in data is measured. Other event-level systematic uncertainties are applied to account for mismodeling of pileup and theory assumptions made during **MC** generation. There are many standard systematic uncertainties derived by **ATLAS**, for example the jet energy measurements or the sagitta measurements of muons, that do not have a large impact on this analysis with its nonstandard physics objects.

L. Horyn, *A Search for Displaced Leptons in the ATLAS Detector*, Springer Theses,
https://doi.org/10.1007/978-3-030-91672-5_11

Table 11.1 Table describing statistical and systematic uncertainties impacting $\tilde{e}$, $\tilde{\mu}$ and $\tilde{\tau}$ efficiencies. Systematics in this table are defined as the difference varying each parameter makes in the final signal yield

Uncertainty source	Uncertainty [%] ($\tilde{e}/\tilde{\mu}$)	Uncertainty [%] ($\tilde{\tau}$)
Statistical	2–46	2–100
Cross section	2–5	2–5
Tracking	2–15	11–14
Muon trigger	1–4	4
Muon selection	3–15	20–37
Electron selection	0.5–2	1–2
Electron trigger	0	0
Lepton displacement	1–18	2–26
Pileup modeling	7	7
Other theory	0–5	0–5
DRAW_RPVLL filter efficiency	1.5	1.5
Luminosity	2	2

11.1 Displaced Lepton Reconstruction

11.1.1 Prompt Lepton Reconstruction

Z bosons can decay into two electrons or two muons. This process is well modeled in **MC**, and easy to identify in data as the invariant mass of the two leptons should equal the Z boson mass (within resolution effects). The invariant mass of the leptons provides a way to identify a lepton as a real lepton that should pass all of the trigger, reconstruction, identification, and selection algorithms. A *tag-and-probe* analysis *tags* events by finding two lepton candidates with invariant mass near the Z boson peak, and then the *probe* leptons are used to measure the selection efficiency by asking that one of the leptons pass a given requirement. This method is used to evaluate trigger, reconstruction, and selection efficiencies of prompt electrons and muons.

The difference between data and **MC** can be thoroughly studied, so scale factors are applied to correct the leptons in MC. A scale factor is defined as:

$$\text{scale factor} = \frac{\text{efficiency in data}}{\text{efficiency in MC}} \tag{11.1}$$

Then the statistical uncertainty on this value is evaluated and applied as an additional uncertainty on the signal. In general, **MC** estimates higher efficiencies than are seen in data. In particular, the **MC** assumes a perfectly aligned detector with all subsystems working perfectly, which is not true in practice, so variables that correlate multiple subdetectors, like electron $\Delta p_{\mathrm{T}}/p_{\mathrm{T}}$ or muon χ^2_{CB}, contribute to a difference between data and **MC**. A 90% scale factor means that **MC** overestimates the selection efficiency by 10%.

ATLAS centrally defines electron and muon scale factors for prompt electrons and muons, but due to the special triggers and selection criteria used in this analysis, custom scale factors are required. For electrons, this analysis uses photon triggers as well as a bug-fixed electron reconstruction, as well as a custom identification and non-standard selection criteria, so all scale factors must be derived specifically for this analysis. For muons, a nonstandard trigger is used, but the reconstruction is standard (except for the tracking, evaluated separately) and the only change to the identification criteria is the removal of the cut on pixel hits, which does not impact prompt muons, so central scale factors are used, with additional selection and special trigger scale factors derived for this analysis.

11.1.1.1 Electrons

For electrons, $Z \rightarrow ee$ events are used to evaluate trigger scale factors, and a single scale factor for reconstruction, identification, and selection as all electrons coming from the Z are real electrons and should pass the trigger as well as all selection criteria.

For both `HLT_2g50_loose` and `HLT_g140_loose` triggers, the efficiency is defined as the number of electrons passing the trigger divided by the number of electrons passing the offline identification criteria. This is done for single electrons, and in the case of the 2 electron trigger, the results are summed in quadrature. It was found that above the trigger threshold, the trigger efficiency in both data and **MC** is very close to 100%, so this scale factor and its associated uncertainty is considered negligible.

The reconstruction, identification, and selection scale factors are evaluated together, by measuring the efficiency for a reconstructed electron candidate to pass the final signal selection. Electron candidates are cluster-track combinations that will get eventually identified as a photon, converted photon, or electron (or none of these). The track reconstruction is studied separately and the cluster reconstruction efficiency at the signal p_T is nearly 100%. The scale factors for electron selection are defined as a function of E_T and η and around 98% except for in the region between the barrel and endcap (around $|\eta| = 1.5$) where it drops to 90%. The statistical uncertainty varies from 1–3%.

11.1.1.2 Muons

$Z \rightarrow \mu\mu$ data and **MC** are used to define additional scale factors for the `HLT_mu60_0eta105_msonly` trigger. The standard trigger scale factors correct for many features of the **MS** and additional corrections are derived and applied on top of them for this analysis. Events are required to pass a **MET** trigger to ensure an unbiased data sample and have two muons within 10 GeV of the mass of the

Z boson. The trigger efficiency is then defined as the number of muons passing the `HLT_mu60_0eta105_msonly` trigger divided by the number of baseline muons. Similarly, selection scale factors are defined by requiring that one muon pass all signal selections (except the $|d_0|$ cut), then the efficiency for the second baseline muon to pass the same signal selection cuts is evaluated.

The scale factors are larger for muons than for electrons because many of the structural features of the **MS** are not well modeled in **MC**. The statistical errors are also around 3%.

11.1.2 Tracking

This analysis relies on **LRT** in order to reconstruct leptons with high p_{T} and high $|d_0|$. Tracks from cosmic muons have high p_{T} and high $|d_0|$ and can be used to measure the tracking efficiency in data. Cosmic muons are tagged as muons with **MS** activity on the other side of the detector, so the cosmic muon must have also passed through the **ID** leaving a track behind. Provided we make some kinematic selections, the existence of the track back-to-back with the cosmic should be solely dependent on the **LRT** efficiency.

A tag-and-probe analysis is performed here as well, by tagging a cosmic muon and looking for tracks back-to-back with it in a narrow ΔR_{cos} cone. Then compare this to a tag-and-probe analysis in signal **MC** by looking for a track in a narrow ΔR cone nearby a truth muon.

There are several important kinematic selections that must be made to ensure the collection of tracks are similar and to correct for the different kinematic distributions between cosmics and data (see Fig. 10.9).

First, to ensure all the hits in the track will be read out with the event, $\phi > 0$ muons must have negative t_0^{avg} (early w.r.t collision), and $\phi < 0$ muons must have positive t_0^{avg} (late w.r.t collision). Making this cut shows a flat reconstruction efficiency w.r.t. t_0^{avg}. All signal muons have very central timing, with **ID** signatures created before **MS** signatures, so this problem does not apply. The impact of the timing cut can be seen in Fig. 11.1.

Second, the cosmic ray muon passes through the detector in an approximately straight line through the **ID**, so the d_0 is simply the distance the cosmic muon was from the PV. This is not the case for signal muons, whose d_0 does not measure a point the muon has gone through, but an extrapolation backwards to the PV. This means that signal tracks with the same d_0 can have very different properties, such as number of hits on track. To correct for this, we require the R_{decay} and d_0 to fall between the same two silicon layers. A sketch of this difference can be seen in Fig. 11.2.

Finally, cosmic muons have a much wider z_0 range than signal muons, which induces an η dependence. So we require cosmic muons to have $|z_0| < 120$ mm, to harmonize with signal muons. Additionally, both signal and cosmic muons must

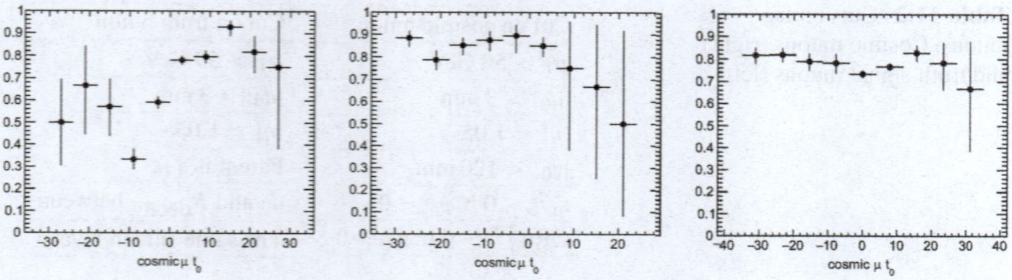

Fig. 11.1 **LRT** efficiency measured with various ϕ and t_0^{avg} cuts. The right shows $\phi < 0$ muons with no t_0^{avg} cut, the center shows $\phi > 0$ muons with no t_0^{avg} cut, and the left shows the final selections, with $\phi > 0$ required to have $t_0^{avg} < 0$ and $\phi < 0$ $t_0^{avg} > 0$. This gives a consistent readout configuration and an approximately flat efficiency w.r.t t_0^{avg}. In particular, there $\phi < 0$ muons with negative timing that result in an artificially low efficiency, likely due to incomplete readout

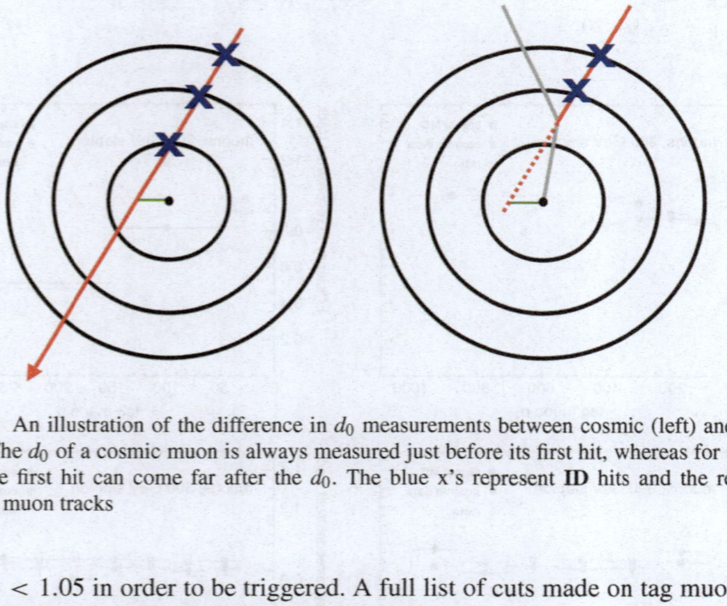

Fig. 11.2 An illustration of the difference in d_0 measurements between cosmic (left) and signal muons. The d_0 of a cosmic muon is always measured just before its first hit, whereas for a signal muon, the first hit can come far after the d_0. The blue x's represent **ID** hits and the red lines represent muon tracks

have $|\eta| < 1.05$ in order to be triggered. A full list of cuts made on tag muons and probe tracks can be found in Tables 11.2 and 11.3.

The remaining difference between cosmic and signal muons is the correlation between p_T and $|d_0|$ in signal. Thus, the cosmic muon p_T distribution is reweighted in each $|d_0|$ bin to match the signal distribution. Then, the ratio of the efficiencies as a function of $|d_0|$ is determined per lepton, shown in Fig. 11.3. The maximum difference, 8%, is taken as the systematic uncertainty per lepton, then summed in quadrature for the two leptons in the event resulting an 11% event-level systematic. Tracking efficiency is assumed to be symmetric around the detector and that after GSF tracking, electron tracking and muon tracking have equivalent efficiency, motivated by Fig. 11.4.

Table 11.2 Cuts on tag muons. Cosmic muons (right) and truth signal muons (left)

Cut on cosmic muon	Cut on truth muon
$p_T > 50$ GeV	$p_T > 50$ GeV
$\|d_0\| > 3$ mm	$\|d_0\| > 3$ mm
$\|\eta\| < 1.05$	$\|\eta\| < 1.05$
$\|z_0\| < 120$ mm	Parent is a $\tilde{\mu}$
$t_0^{avg} > 0$ if $\phi_\mu < 0$	d_0 and R_{decay} between
OR $t_0^{avg} > 0$ if $\phi_\mu > 0$	The same silicon layers

Table 11.3 Cuts on probe **ID** tracks. In data (right) and signal (left)

Cut on cosmic muon	Cut on truth muon
$p_T > 30$ GeV	$p_T > 30$ GeV
$\Delta R_{\cos} = \sqrt{(\Delta\phi - \pi)^2 + (\Sigma\eta)^2} < 0.3$	$\Delta R = \sqrt{(\Delta\phi)^2 + (\Delta\eta)^2} < 0.05$
$\|d_{0,\text{track}} - d_{0,\mu}\| < 20$	
$\|z_{0,\text{track}} - z_{0,\mu}\| < 20$	

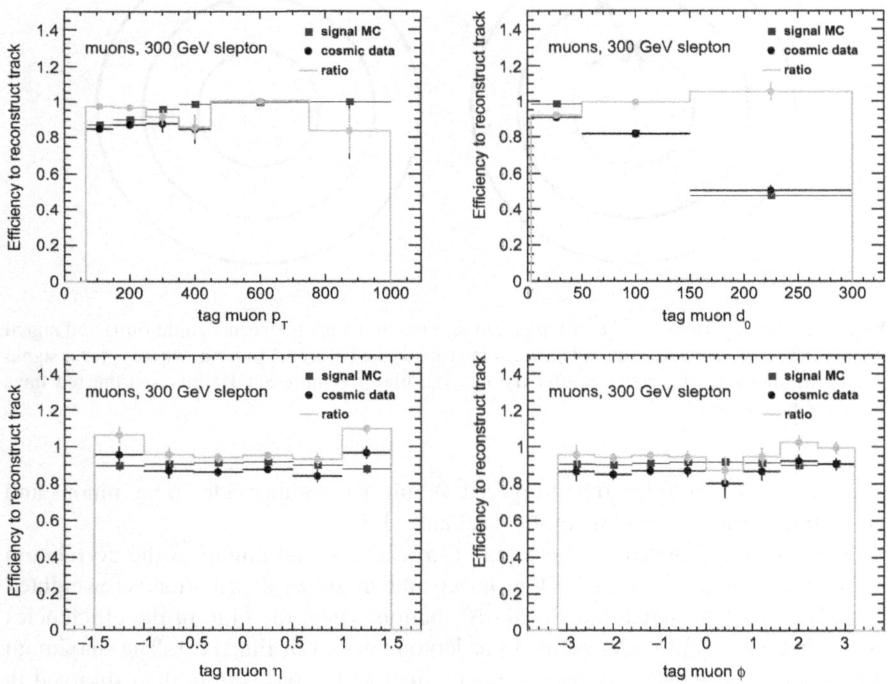

Fig. 11.3 A comparison of tracking efficiency in cosmics data and signal MC (all masses and lifetimes can be used to due the eventual p_T and d_0 binning) with respect to p_T (top left), d_0 (top right), η (bottom left), and ϕ (bottom right). The **MC** efficiency is shown in purple, the data efficiency in black, and the ratio between the efficiencies is shown in gray

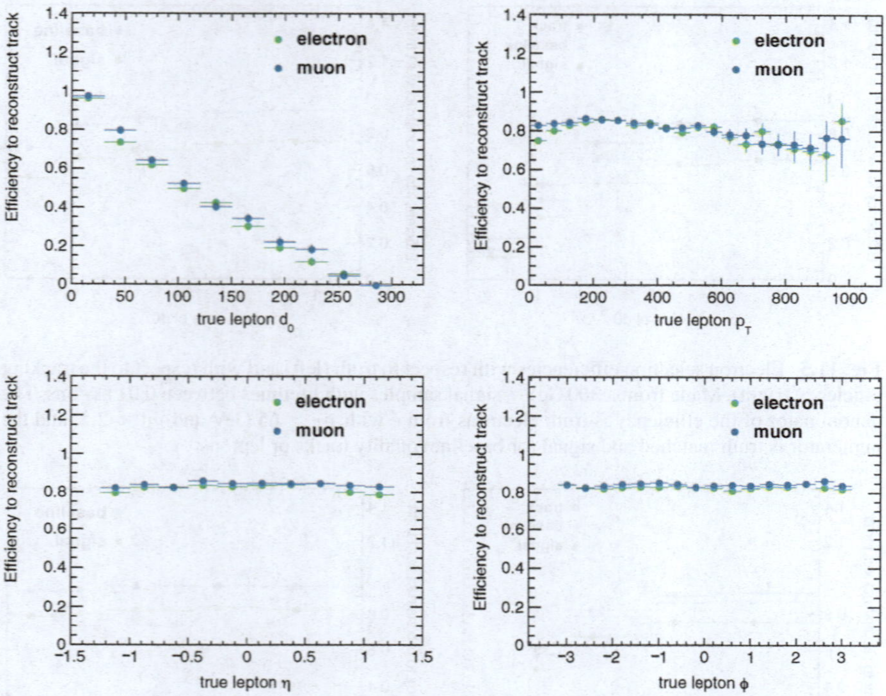

Fig. 11.4 Tracking efficiency for electrons and muons in signal MC (all lifetimes of 300 GeV $\tilde{e}$ or $\tilde{\mu}$). These plots justify the assumption that tracking efficiency is the same between electrons and muons and symmetric about the **ID** volume

11.1.3 Lepton Displacement

The effect of displacement on lepton reconstruction cannot be directly compared between data and **MC**. An additional uncertainty is defined in **MC** to account for reconstruction effects potentially not captured by the previous two systematic uncertainties. It is taken as the deviation in the reconstruction efficiency as a function of $|d_0|$ relative to prompt muons. This uncertainty is meant to capture any discrepancy in **MS** track or electron shower shape parameters at high $|d_0|$, as well as changes in how they are combined with the **ID** track. This uncertainty is extremely conservative as there is no way to determine if the same fluctuations are seen in data.

To determine the systematic uncertainty due to displaced reconstruction, the baseline or signal reconstruction/selection efficiency is divided by the tracking efficiency in order to separate any inefficiencies from **LRT** shown in Figs. 11.5 and 11.6. Then, the ratio of each high d_0 bin relative the prompt bin (0–3 mm) is taken, results of this are shown in Fig. 11.7. The uncertainty is assigned to each lepton and they are summed in quadrature to get an event level systematic.

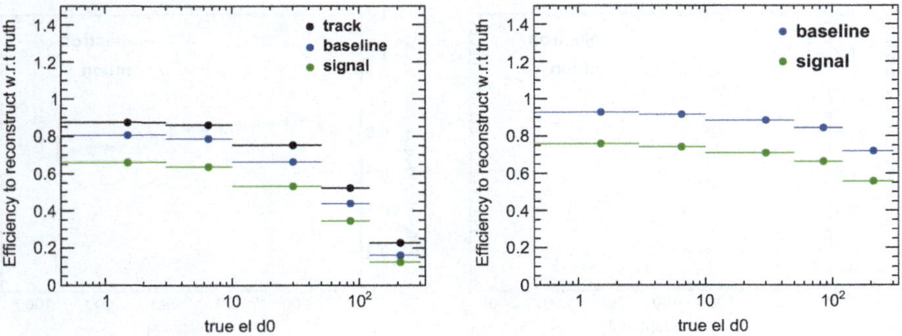

Fig. 11.5 Electron selection efficiencies with respect to truth (left) and with respect to the tracking efficiency (right). Made from a 300 GeV $\tilde{\ell}$ signal samples with lifetimes between 0.01 ns–1 ns. The denominator of the efficiency is truth electrons from $\tilde{e}$ with $p_{\mathrm{T}} > 65$ GeV and $|\eta| < 2.5$, and the numerator is truth matched and signal (or baseline) quality tracks or leptons

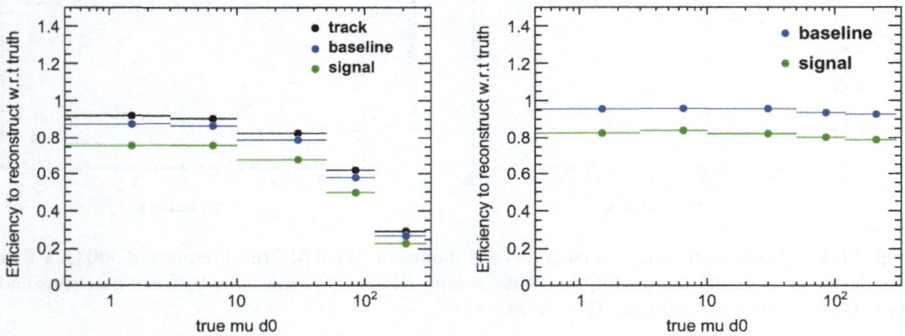

Fig. 11.6 Muon selection efficiencies with respect to truth (left) and with respect to the tracking efficiency (right). Made from 300 GeV $\tilde{\ell}$ signal samples with lifetimes between 0.01 ns–1 ns. The denominator of the efficiency is truth muons from $\tilde{\mu}$ with $p_{\mathrm{T}} > 65$ GeV and $|\eta| < 2.5$, and the numerator is truth matched and signal (or baseline) quality tracks or leptons

This uncertainty for muons is quite small, while it is much more substantial for electrons. Electrons are identified using a likelihood, which is trained with d_0 as a discriminating variable. $|d_0|$ was removed from the cuts performed offline, but the LH was not retrained, and so there as a larger relic d_0 dependence and the displacement uncertainty is much larger.

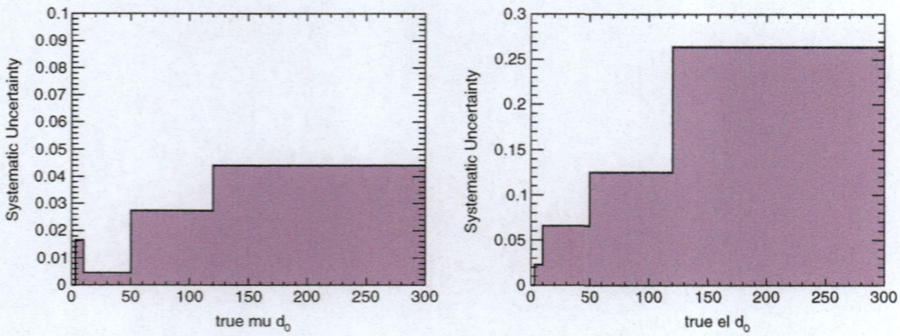

Fig. 11.7 Fractional systematic uncertainties defined for muons (left) and electrons. The value of each bin is defined as 1 minus the ratio of the selection efficiency with respect to tracking efficiency of the given bin to the same value of the prompt (0–3 mm) bin. These are defined in 300 GeV $\tilde{\ell}$ signal samples with lifetimes between 0.01 ns–1 ns. It was confirmed that the trends are consistent across various $\tilde{\ell}$ masses

11.1.4 Other Sources of Uncertainty

11.1.4.1 Pileup Modeling

When **MC** is generated, particularly when it is generated during the course of the run, the actual pileup distribution of the events from the **LHC** is not known. This is corrected through a process called *pileup reweighting*, where a more realistic pileup profile is added to **MC** events. The change in number of signal events when the pileup profile is varied is taken as a systematic uncertainty, 2%.

11.1.4.2 Theoretical Uncertainties

Additional uncertainties are taken for the renormalization and factorization scales that are used to generate the physical processes in **MC**. These impact both the cross section measurement and the final lepton kinematics. Both scales are varied, impact on the final results quantified, and the range of variation is taken as an uncertainty, in this analysis about 5%.

11.1.4.3 Luminosity Measurement

ATLAS measures luminosity using dedicated detectors and calibrations (discussed in Sect. 3.4. The uncertainty on this measurement contributes a 2% uncertainty to the analysis, as it impacts the normalization of the signal yield from **MC**.

Part V
Results

Chapter 12
Results

A search for leptons with large impact parameters in 139 fb^{-1} of pp data with $\sqrt{s} = 13$ TeV collected by the ATLAS detector between 2015 and 2018 has been presented. Three Signal Regions (SR) are defined based on the flavor of the two highest p_T leptons in the event: SR-ee with two electrons, SR-$\mu\mu$ with two muons, and SR-$e\mu$ with an electron and a muon. Quality requirements are designed to select real leptons with large impact parameters. Backgrounds from algorithm fakes, heavy-flavor meson decays, and cosmic-ray muons are estimated from data.

12.1 Signal Yield

Zero events are observed in each signal region, in full agreement with the background estimate. The result of each cut made in each **SR** can be seen in Tables 12.1, 12.2, and 12.3 for SR-$\mu\mu$, SR-ee, and SR-$e\mu$, respectively. The most powerful background discriminators are the $\Delta p_T / p_T$ cut for electrons and cosmic veto for muons, also in agreement with expectations.

This analysis was designed to be as general as possible, so the results can be applied to any new physics signature that results in 2 leptons with high $|d_0|$ and high p_T.

L. Horyn, *A Search for Displaced Leptons in the ATLAS Detector*, Springer Theses,
https://doi.org/10.1007/978-3-030-91672-5_12

Table 12.1 Cutflow for SR-*ee* for Run 2 data

Cut	Data yield
Pass trigger and at least 2 baseline leptons	65,484
2 leading leptons are electrons	19,419
$p_T > 65$ GeV	14,061
$\lvert d_0 \rvert > 3$ mm	11,589
Both electrons pass isolation	9220
$\Delta p_T / p_T \geq -0.5$	7
$\chi^2_{ID} < 2$	2
$N_{miss} \leq 1$	0
$\Delta R_{\ell\ell} > 0.2$	0

Table 12.2 Cutflow for SR-$\mu\mu$ for Run 2 data

Cut	Data yield
Pass trigger and at least 2 baseline leptons	65,484
2 leading leptons are muons	45,845
$p_T > 65$ GeV	35,607
$\lvert d_0 \rvert > 3$ mm	35,326
Both muons pass isolation	35,204
Muons pass cosmic veto	2
$t_0^{avg} < 30$	2
$\chi^2_{ID} < 2$	2
$N_{miss} \leq 1$	2
$N_{prec} \geq 3$	0
$\chi^2_{CB} < 3$	0
$N_\phi > 0$	0
$\Delta R_{\ell\ell} > 0.2$	0

Table 12.3 Cutflow for SR-$e\mu$ for Run 2 data

Cut	Data yield
Pass trigger and at least 2 baseline leptons	6548
2 leading leptons are a muon and an electron	1910
$p_{\mathrm{T}} > 65\,\mathrm{GeV}$	128
$\lvert d_0 \rvert > 3\,\mathrm{mm}$	107
Both leptons pass isolation	98
Muon pass cosmic veto	86
Muon $t_0^{\mathrm{avg}} < 30$	82
Muon $\chi^2_{\mathrm{ID}} < 2$	72
Muon $N_{\mathrm{miss}} \leq 1$	68
Muon $N_{\mathrm{prec}} \geq 3$	6
Muon $\chi^2_{\mathrm{CB}} < 3$	0
Muon $N_{\phi} > 0$	0
Electron $\Delta p_{\mathrm{T}}/p_{\mathrm{T}} \geq -0.5$	0
Electron $\chi^2_{\mathrm{ID}} < 2$	0
Electron $N_{\mathrm{miss}} \leq 1$	0
$\Delta R_{\ell\ell} > 0.2$	0

Chapter 13
Interpretation

Since no events are observed, limits are set on the parameter space of **GMSB SUSY** models. Limits are computed using HistFitter [1], an **ATLAS** framework that combines the observed number of events with uncertainties on background predictions and signal systematics and calculates both model dependent and model independent limits using the $\mathrm{CL_s}$ technique [2]. $\mathrm{CL_s}$ is a useful tool for particle physics analyses. Somewhere between a purely frequentist and purely Bayesian method of deriving a confidence interval, the $\mathrm{CL_s}$ describes the confidence in a signal-only hypothesis. The limit curve is drawn where $\mathrm{CL_s}(m_{\tilde{\ell}}) \leq 5\%$, meaning that the probability of having falsely excluded a $\tilde{\ell}$ of a given mass is less than or equal to 5%. The smaller the value of $\mathrm{CL_s}$, the lower the probability that a $\tilde{\ell}$ with a given mass exists. For a search where 0 background events are predicted, 3 events are required for a 95% $\mathrm{CL_s}$ [3].

13.1 Slepton Limits

Limits are set on the possible masses and lifetimes of long-lived $\tilde{\ell}$. Four different limits are set using the results of this analysis: each $\tilde{e}$, $\tilde{\mu}$, or $\tilde{\tau}$ **NLSP**, or the mass degenerate case with all three as co-**NLSP**s. Results for all four scenarios combined are shown in Fig. 13.1, and the individual limit for each of the four scenarios along with their observed $\mathrm{CL_s}$ value is shown in Fig. 13.2.

For the $\tilde{e}$ **NLSP** and $\tilde{\mu}$ **NLSP** cases, the results from SR-ee and SR-$\mu\mu$, respectively, are used. While for the $\tilde{\tau}$ and co-**NLSP** cases, all three SRs are combined. For a lifetime of 0.1 ns, $\tilde{e}$ NLSP, $\tilde{\mu}$ **NLSP**, $\tilde{\tau}$ **NLSP**, and co-NLSP scenarios are excluded for $\tilde{\ell}$ masses up to 720 GeV, 680 GeV, 340 GeV, and 820 GeV, respectively. Co-NSLP events are also excluded up to 10 ns for masses below 330 GeV. The previous from OPAL [4] is surpassed by nearly an order of magnitude for $\tilde{e}$-, $\tilde{\mu}$-, and co-NSLP scenarios.

L. Horyn, *A Search for Displaced Leptons in the ATLAS Detector*, Springer Theses,
https://doi.org/10.1007/978-3-030-91672-5_13

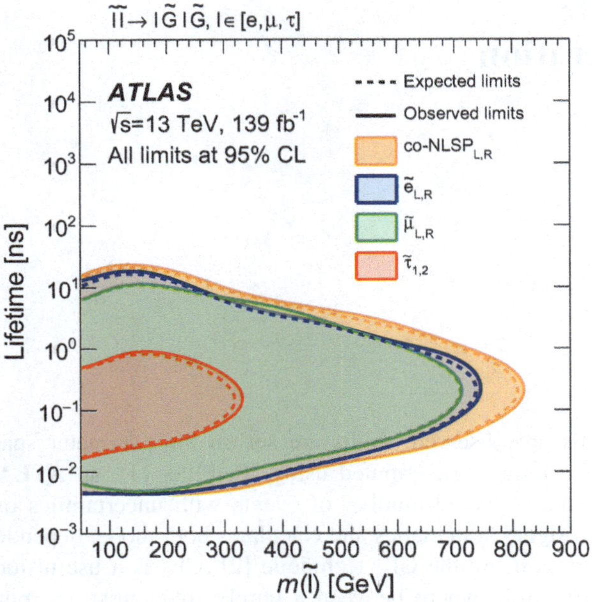

Fig. 13.1 Expected (dashed) and observed (solid) exclusion contours for $\tilde{e}$ **NLSP**, $\tilde{\mu}$ **NLSP**, $\tilde{\tau}$ **NLSP**, and co-NLSP production as a function of the lifetime at 95% CL

13.1.1 Comparison to LEP

The summary in Fig. 13.1 includes both right- and left-handed slepton production and cannot be directly compared to the results from **LEP**. Right-handed only or left-handed only, as well as $\tilde{\tau}_1$ and $\tilde{\tau}_2$ limits are shown in Fig. 13.3 and compared to the LEP limit where possible.

13.2 Model-Independent Limits

Additionally, model independent limits are set on generic new physics processes in the **SRs**. These limits are based on the visible cross-section of new physics ($\langle\epsilon\sigma^{95}_{\text{obs}}\rangle$) and the observed ($S^{95}_{\text{obs}}$) and expected ($S^{95}_{\text{exp}}$) number of signal events that would be measured. Visible **BSM** cross sections above 0.02 fb are excluded in each **SR** These values are shown in Table 13.1

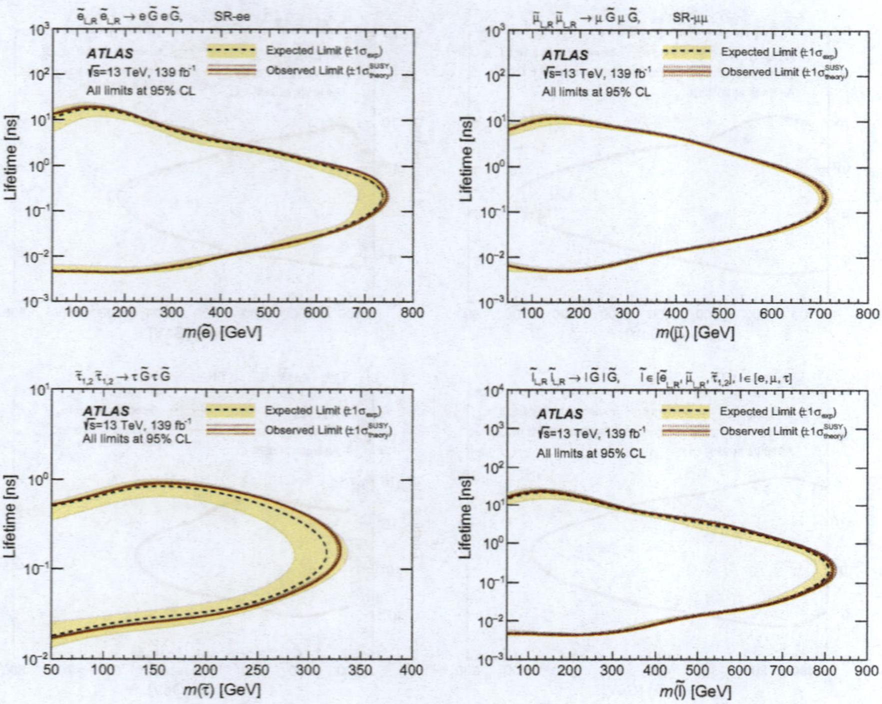

Fig. 13.2 Individual exclusion curves for four different **NLSP** scenarios: $\tilde{e}$ (top left), $\tilde{\mu}$ (top right), $\tilde{\tau}$ (bottom left), co-**NLSP** (bottom right)

Table 13.1 Left to right: 95% CL upper limits on the visible cross section ($\langle\epsilon\sigma\rangle^{95}_{\text{obs}}$) and on the number of signal events (S^{95}_{obs}). The third column (S^{95}_{exp}) shows the 95% CL upper limit on the number of signal events, given the expected number of background events. The last column shows the p-value of the no-signal hypothesis

Signal channel	$\langle\epsilon\sigma\rangle^{95}_{\text{obs}}$[fb]	S^{95}_{obs}	S^{95}_{exp}
SR-ee	0.02	3.0	3.0
SR-$\mu\mu$	0.02	3.0	3.0
SR-$e\mu$	0.02	2.7	3.0

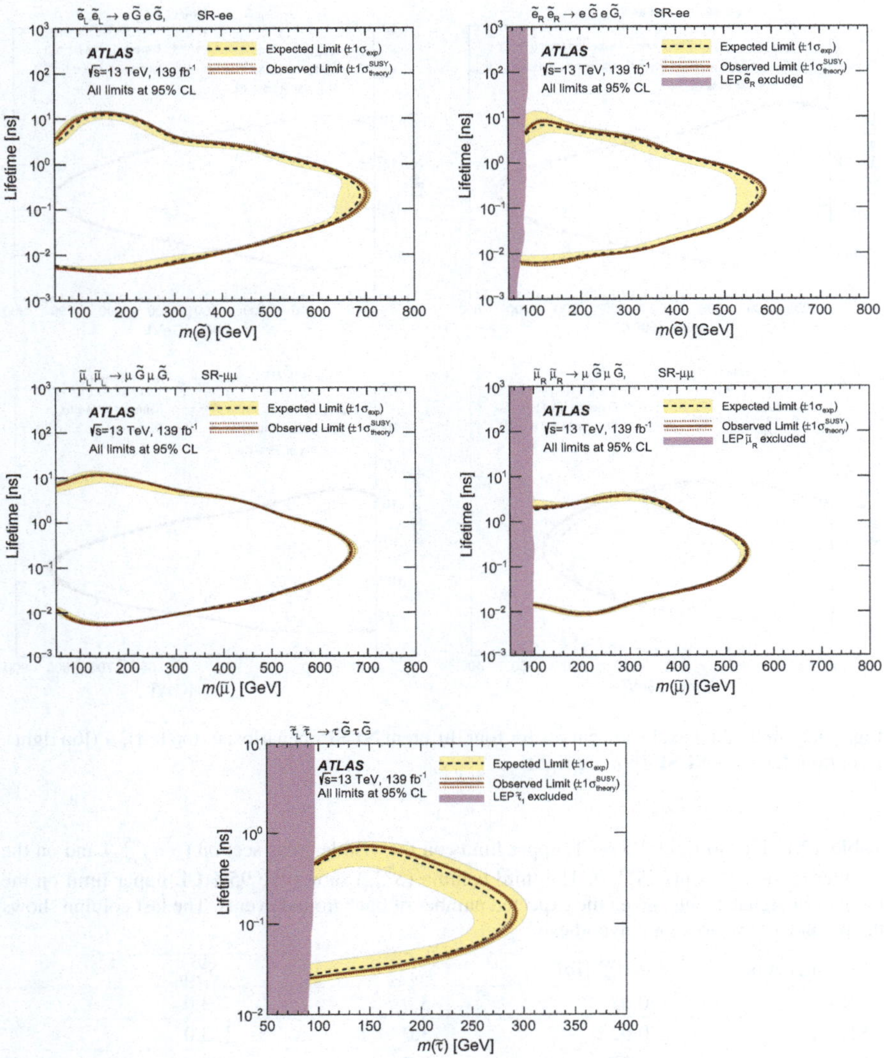

Fig. 13.3 Limits on $\tilde{\ell}$ separated by their chirality with the **LEP** limit shown in pink. In order from top to bottom, left to right: $\tilde{e}_L$, $\tilde{e}_R$, $\tilde{\mu}_L$, $\tilde{\mu}_R$, left-handed $\tilde{\tau}_1$. This result does not have significant sensitivity to right-handed $\tilde{\tau}_2$

References

1. M. Baak et al., HistFitter software framework for statistical data analysis. Eur. Phys. J. C **75**(4), 153 (2015). https://doi.org/10.1140/epjc/s10052-015-3327-7.
2. A.L. Read, Presentation of search results: the CL(s) technique. J. Phys. *G* **28**, 2693–2704 (2002). https://doi.org/10.1088/0954-3899/28/10/313
3. K. Bjoerke et al., Summary of and Recommendations for the "Better than Zero" Problem. Technical report. ATL-COM-GEN-2017-009. Etai Nativ is another contributor to this note, but cds won't let me enter his name. GRRR!! Geneva: CERN (2017). https://cds.cern.ch/record/2280679
4. The OPAL Collaboration, Searches for Gauge-Mediated Supersymmetry Breaking topologies in e^+e^- collisions at centre-of-mass energies up to $\sqrt{s} = 209$ GeV. Eur. Phys. J. C **46**(2), 307–341 (2006). https://doi.org/10.1140/epjc/s2006-02524-8

Chapter 14
Conclusions and Future Improvements

The lifetime is an underexplored region of phase space for many natural **BSM** theories and this thesis describes a search that covers a large gap in signature space at the **LHC**. In 139 fb^{-1} of $\sqrt{s} = 13$ TeV **LHC** data collected by the **ATLAS** detector, no signs of **BSM** physics are seen. Three orthogonal signal regions are defined and less than 1 background event is predicted in each. Zero events are seen and limits are set on the possible mass and lifetime of $\tilde{\ell}$ in **GMSB SUSY** models. For a lifetime of 0.1 ns, selectron NLSP, smuon NSLP, stau NSLP, and co-NLSP scenarios are excluded for slepton masses up to 720 GeV, 720 GeV, 370 GeV, and 830 GeV, respectively, exceeding the LEP co-NSLP limit [1] by almost an order of magnitude. Co-NSLP events are also excluded up to 10 ns for masses below 330 GeV.

A minimal, though substantial, set of optimizations were done to obtain this result, but future analyses using the same dataset could improve upon this result in several ways. First, further optimization could be done of the electron reconstruction algorithm in order to boost efficiency at high $|d_0|$ as well as reduce the systematic uncertainty due to the variation in the lepton displacement selection efficiency. Also, additional signal regions could be defined using a **MET** trigger to increase sensitivity. This signal model has real **MET** due to the gravitinos (and neutrinos in the $\tilde{\tau}$ decays). More importantly, muons are not included in the **MET** calculation at the trigger stage, so an **MET** trigger is an effective displaced muon trigger. A bolder improvement would be to reconstruct all recorded events with **LRT**, instead of the filtered 10% that is currently used. This would enable more creativity in signal region design and more centralized displaced lepton identification algorithms.

Additional optimizations should be performed to prepare for an analysis in Run 3. In particular, this search suffered from trigger limitations. A simple improvement to the data collection scheme would be to introduce an EM-only + MS-only trigger. This would likely have a low enough fake rate that it could be added in relatively easily and p_{T} and η requirements on the single EM-only and MS-only signatures could be relaxed. A more exciting improvement would be to implement **LRT** in

L. Horyn, *A Search for Displaced Leptons in the ATLAS Detector*, Springer Theses,
https://doi.org/10.1007/978-3-030-91672-5_14

the trigger and design trigger-level displaced lepton identification algorithms. This would likely have a low enough rate to reduce the p_T requirements substantially.

This result should also be interpreted in conjunction with other searches, including prompt searches and those for disappearing tracks and stable massive particles, to probe the full possible lifetime space of sleptons and make a more lifetime-inclusive statement about **GMSB SUSY** at the **LHC**. Furthermore, the minimal event-level requirements make this result model-independent and applicable to any **BSM** decay resulting in displaced leptons.

Reference

1. ALEPH, DELPHI, L3, OPAL Experiments. Combined LEP Selectron/Smuon/Stau Results, 183-208 GeV. LEPSUSYWG/04-01.1 (2004). http://lepsusy.web.cern.ch/lepsusy/www/sleptons_summer04/slep_final.html